SOLARIAN LEGACY

SOLARIAN LEGACY

Metascience and a New Renaissance

by

Paul Von Ward

Inner Eye Books

An Imprint of

OUGHTEN HOUSE PUBLICATIONS

Livermore, California USA

SOLARIAN LEGACY: Metascience and a New Renaissance
by
Paul Von Ward

Published in 1998

00 99 98 97   0 9 8 7 6 5 4 3 2 1

Published by
Inner Eye Books
an imprint of
OUGHTEN HOUSE PUBLICATIONS
PO Box 2008
LIVERMORE, CA 94551
PHONE: (510) 447-2332
FAX: (510) 447-2376
E-MAIL: oughten@oughtenhouse.com
INTERNET: www.oughtenhouse.com

Library of Congress Cataloging-in-Publication Data

Von Ward, Paul, 1939-
Solarian Legacy : metascience and a new renaissance / Paul Von Ward.
p. cm.
Includes bibliographical references and index.
ISBN 1-880666-75-8
1. Microcosm and macrocosm--Miscellanea. 2. Science--Miscellanea. 3. Spiritual life--Miscellanea. I. Title.
BF1999.V66 1997
113--dc21 97-30460
CIP

Printed in USA

For

Sandra Laurissa, Michael David, and Stephen Paul,

who will live to fully grasp the import of
their Solarian Legacy.

ACKNOWLEDGMENTS

Numerous people have contributed to this book: it is impossible to individually recognize them all. Many friends and colleagues have shared their ideas and given personal reactions to the concepts described here. I hope they recognize their thoughts. If my perspective distorted them, I alone am responsible for the manner in which I responded to their contributions.

During the early days of writing, I was borne up by the energetic encouragement and technical support of Brenda Sanchez and the freedom provided by the job-sharing with Tonette Long in our co-leadership of Delphi International. Both generously gave of their ideas, time, and energy to my efforts.

Among the many who have given of their time for review and feedback on the manuscript, or parts of it, I particularly express my appreciation to Nancy Parker, Rosemary McMullen, Celeste White, Michael Brein, Miriam Saadi, Phil Metlzer, Brian O'Leary, Alicia Mannix, Mariah and Kristos, Elliott Dacher, and Elsa Porter. I also thank the many unnamed others who have reacted to specific ideas and given me the benefit of their experience.

Special thanks go to my two dedicated editors, Judith Cope and Tonette Long, without whose professional skill and personal sensitivities this text would not have materialized. The two, however, are not responsible for any of its deficiencies; they result from my own judgments.

My appreciation is given to Rosemary Whitford of InForms Publications and Jim Schuette and Kelesyn Winter of Wellspring Publishing for their technical support, graphic design, and layout, with Kelesyn being responsible for the excellent graphics and Jim for the comprehensive index. Laura Kay and Don Kay are responsible for the cover design. And the photograph on page 304 is courtesy of the late Roy Lee.

And last, but not least, I must thank Milton Jetter who, in one moment of synchronicity, gifted me a copy of *The Kybalion*, with its exposition on the Hermetic Principles, just as I was grappling with the structure of the book to come.

CONTENTS

Preface

During more than three decades of professional study, field research, and practical experience living in a number of different cultures, I became increasingly perplexed as to why individuals and institutions continued to hold dearly beliefs that were not consistent with established fact and were contrary to human experience. This quandary was only deepened by my successive involvement in religious, academic, military, diplomatic, government, and business institutions. Individuals who raised questions about the incongruencies between organizational assumptions and experienced reality usually found themselves at odds with the institutions. Instead of being able to contribute to internal reform, most were forced to leave.

This apparent natural resistance to the imperative to change was dealt with in my first book of more than fifteen years ago. *Dismantling the Pyramid: Government by the People* was an effort to develop principles conducive to self-renewing government organizations. Ideas like some of those in it contributed to the subsequent U.S. effort known as "reinventing government," which has been no more successful in effecting profound change than any number of previous Washington-based reform initiatives. Over the last decade, I have realized the problem is much more fundamental than changing organizational procedures and structures. The challenge of remaining open to learning from experience and making changes that lead to the realization of greater potential calls into question the basic world views that are essential to holding the human psyche together. To be able to ride the spiral of continuing self-renewal and to increase one's capacity for experiencing more complex facets of life requires conscious attention to one's definition of self.

My experience is that most do not choose voluntarily to wrestle with long-established definitions of their realities. However, from time to time, individuals and societies are given reason to pause, assess their performance, and decide if they are living up to their potential. Circumstances—economic, political, natural, or personal disasters—may lead them to conclude course corrections are necessary. In such times they may dare to question long-held, life-shaping shibboleths. In the midst of current family, institutional, and societal problems, I believe large numbers of Americans, and significant percentages of the world's peoples, are now ready to question the veracity of some inherited myths that underlie modern society. Many are ready to reexamine their heritage, to consider whether they are taking full advantage of their human legacy.

Therefore, over the last five years, in the company of some independent scientists and metaphysicians, I have engaged in an assessment of many of society's most cherished assumptions about the history of the Earth and its human inhabitants. Published research, discussions in this community of independent scholars, and personal testing have dramatically reshaped my perspective on the human legacy and, consequently, even my approach to daily life.

As humanity faces some of the most profound challenges in its history, I believe it is simultaneously provided with the internal resources to meet them. This book and others with similar material expand the reader's view of this universe, and bring humanity closer to the truth of its inner nature. Though we cannot yet see beyond the curve of space-time, we might open our eyes to a fuller spectrum of our multidimensional reality. Taking just some preliminary steps reveals that there is much more to the human experience than we normally like to admit.

The book that follows articulates a provisional, but more realistic picture of the human legacy by drawing together the ideas of countless pioneers who have shared a willingness to test alternative perceptions of reality. Various chapters examine material that has been either dismissed as unscientific by the intellectual establishment or ignored because of its unconventional sources. Included are the improperly labeled "paranormal" phenomena, reported anomalies of intelligent nonhuman forms, intriguing evidence of forgotten advanced civilizations, compelling findings of frontier science, and wisdom germinated in some cosmic seedbed that springs up in unexpected ways. Though the material and its implications can be disturbing at first encounter, I believe it grounds us in a more stable field of conscious awareness. I try to balance openness with a healthy skepticism in this effort to nudge readers beyond a period of self-imposed blinders, into an era of *self-conscious cosmic beings*.

The intent is not to present a fully documented and exhaustively researched case for academic peer review. Though the book is not designed to fit the research protocols of any scientific discipline, I hope it speaks to the specialist as well as the inquisitive lay reader.

Although I am passionate about the ideas set forth here, my fervor should be tempered by the reader's own experience. As this book reflects my ongoing testing, I invite readers to put the explicit and implicit assumptions to critical review. They should be found credible only if the following conditions are met:

- The integration of my ideas makes intellectual sense to the reader
- The interpretations resonate with the reader's own observations, and
- The reader's own experiences validate the hypotheses.

While I have attempted to make the material in the text accessible to all readers, some sections may go into more depth than is of value to a particular reader. For example, if one has less interest in the details of astronomy, physics, and chemistry discussed here, parts of the first two chapters may be skimmed without missing the main themes. However, the discussion of the "void" in Chapter 2 is an essential foundation for later chapters. For those who are well versed in the history of the UFO phenomenon and alternative views of human prehistory, parts of Chapter 3 can be passed over lightly. Because it sets forth significantly different ideas about the nature and scope of science and human capacities, the somewhat complex Chapter 4 is a must read for understanding the remainder of the book. Chapters 5 and 6 present a new perspective on the role of consciousness in the universe and its incarnation in individual beings. Part III is an exploration of that perspective in terms of human experience, and the potential for a New Renaissance.

In the final analysis, readers will take these ideas seriously not because the author is a "chosen person" with a "special message" from "the source of knowledge or power," but because they recognize a fellow seeker involved in the universal experience of conscious life.

Paul Von Ward
Ashland, Oregon

Introduction

Our highest endeavor must be to develop free human beings, who are able of themselves to impart purpose and direction to their lives.
Rudolph Steiner

This book is a prologue to a possible new planetary myth, a revised legacy for the human race. Using the best of our current knowledge, it seeks to correct parts of the conventional story we tell about ourselves and to set the stage for a more truthfully lived future.

Most academic scholarship provides the general public with a rather limited sense of planetary history and humanity's place in it. Anything more than ten thousand years ago is considered prehistory. Views of civilization held by most Westerners are focused on less than the last five thousand years, starting in the Middle East and spreading from Mesopotamia and Egypt through Greece and the Roman Empire to the rest of the world. Scholars are not quite sure why things seemed to flourish so quickly in these regions, how preliterate tribes developed sophisticated agricultural techniques and complex cities without obvious antecedents. Learning from research that 65,000 years ago Cro-Magnon man in Europe had a brain capacity equal to twentieth century man's, they assume that most of those years were spent in primitive hunting and gathering activities. This brain development is seen as the result of millions of years of biological evolution, which will be understood as soon as someone finds the "missing link." Other alternatives are not taken seriously.

This over-simplification of history, recently portrayed in the popular *Newsweek* (March 24, 1997) conventional view of life at the time of the previous passage of the Hale-Bopp comet, omits mounds of evidence pointing to more ancient civilizations, including prehistoric evidence of advanced intelligence in remains of vast cities and sophisticated artifacts. Scientists are equally culpable, offering no explanations for large sectors of human experience. Just as serious scholars cannot ignore physical ruins and archaeological finds that are left out of history books, objective scientists cannot pretend that anomalous human experiences do not exist simply because they are not part of the current scientific paradigm.

The following chapters are intended to challenge several assumptions inherent in the conventional view of the human heritage. They provide a basis for a less biased description of the human species' history and its inherited resources that are labeled here the "Solarian Legacy." While reviewing relevant evidence for the Solarian Legacy, readers are encouraged to consider adding the following perspectives to their personal assumptions about the human passage into the twenty-first century:

- Intelligent human life has existed on Earth much longer than most believe
- Science and technology at least as advanced as that of the twentieth century were achieved millennia ago
- The history of Earth and its inhabitants has been dramatically affected by cataclysmic events that leave gaps in the archaeological record
- Human beings are more mentally, psychologically, and physically interactive with each other and the cosmos than we imagine
- Our consciousness is not limited by the world of five senses
- Human beings have much more inner power to influence internal and external events than most now understand
- Humans coexist with other highly intelligent beings and have a long history of interaction with them
- All conscious beings function in a multidimensional reality
- Humanity is on the threshold of understanding its true identity.

While the above elements of our Solarian Legacy are not intended to comprise the totality of the truth, they do suggest reasonable working hypotheses to fill the gaps in "official knowledge" addressed in this book. These hypotheses suggest an alternative perspective on large chunks of our intellectual heritage, including academic disciplines that underpin popular concepts of human development, social institutions, political and economic systems, as well as biological and physical sciences. They call into question core tenets of religions, psychological theories, beliefs about instincts and motives, concepts of male and female, the nature of intelligence, in effect, the current definition of human identity. They require reexamination of ethical and value systems based on obedience to divine power or regionally defined gods. Consideration of these hypotheses requires reviewing all appeals to a historically-based authority. A new definition of our heritage is called for, one that recovers our complete history and the intrinsic links between spirit and matter. Then the foundation can be laid on which to build a new global consensus on principles for planetary harmony.

Widespread engagement with the larger historical reality being uncovered by recent discoveries will correct our myopia about our past and potential. Expanding explorations in the realms of extraordinary or inner reality and the growing acceptance of evidence of other conscious beings in our space-time give us an entirely new sense of who we are. We are beginning to see humanity and its potential as a part of varying systems or levels of reality embedded in and superimposed on each other.

The term "isomorphism" applies to two systems that, except for one variable, behave according to the same principles. For example, in hydraulics and aerodynamics, all the laws that apply to one apply to the other, though the former deals with fluids and the latter, with air. A similar isomorphism applies in the fields of ecology and consciousness research. What we have learned about the interrelated and indivisible nature of ecological systems applies equally to the study of cosmic consciousness. While developing an understanding of how our physical pollutants affect the whole system, we have yet to grasp that our spiritual, mental, and psychological pollutants contaminate the global society and the cosmos beyond. These personal pollutants engender distress for others, and return to darken our own souls.

In a unitary system there are no strictly local effects: perturbations in any person or group reverberate through the whole. As ecoethics remind us of our personal responsibility for the health of the whole planet, we need an ethical code for conscious living that holds our focus on the long-range implications for others, and for ourselves, of our thoughts, words, and deeds. Because the full physical, emotional, and intellectual development of any one being is dependent on the progress made by all, no individual salvation or one-by-one escape to nirvana exists: cosmic self-realization must occur, paradoxically, collectively as well as individually. But a guiding, conscious image of that singular reality now eludes most of us.

People feel fragmented and disconnected from their species and the planet at large: there is growing recognition that important bonds are missing from modern life. Today's technological breakthroughs, institutional fragmentation, and social complexity have rendered the practical business of daily living a seemingly superficial experience, disconnecting us from the inner dimensions of group life and selfhood. Rites of passage no longer unite us within and among generations. Our emotional needs for passionate group experiences of celebration and despair, in birth, life, and death, go unfilled. People are left alone to create idiosyncratic interpretations for seemingly disparate events, and life thereby loses much of its cohesion and meaning.

Academics and scholars among us divide reality into smaller and smaller bits, and write and teach more and more about less and less. Those of us who are scientists study thinner and thinner slices of life and forget what a person, plant, or planet looks and feels like. Religious leaders expend enormous time and energy distinguishing their version of reality from that of other religions, and defaming those they fear would steal their flocks. Politicians let partisan concerns distract them from the overall good. The result is social and emotional chaos, an era of transition that calls for a new sense of direction.

Centuries ago the two great intellectual forces in modern life, science and religion, left their common roots and branched onto separate paths. Each in its own way has been severed from the inner experience of emotion and intellect that continually shape the individual's life experience. Yet within each person there is an inherent impulse to return to a unified state of being, a oneness with the whole. People sense themselves as part of something inexplicably vast when holding the hands of dying loved ones who slip in and out of other realms. Farmers and botanists among us witness an inner order in dry seeds that, in answer to some secret bidding, spring from the soil and grow into the plants that serve our needs for food and beauty. Sensing the pull of the stars, we seek to understand the influences of forces beyond our planet.

But why have we lost sight of our interdependence with each other, nature, and the stars? Why have we been so busy dividing the territory among ourselves that we lose sight of our interconnectedness and fail to treat each person as part of the whole? Desiring to stake out areas of specialized knowledge and control, why do we lose sight of the cosmos from which every being arises? These questions point to our *need for a better vision of the human place in the larger order, one that fits reality more accurately than the current myths about our past and potential.*

A comprehensive new picture is within the grasp of all conscious beings. An intellectual elite is not necessary to interpret the fundamentals of the universe for the masses. *No true and perennial wisdom is arcane*; only those who wish to exercise power over others would disagree. Each person has the capacity to grapple with the three existential questions: From where do I come? What is my purpose? Where do I go after this? In each cycle of waking and sleeping, both of which are filled with consciousness, life gives us enough data to test postulated answers to these questions. As a cosmic citizen, *each of us is capable of being touched by and of touching the farthest reaches of the universe.*

The new unifying picture must include a more complex sense of reality, portraying the fullness of physical, emotional, and conscious experience. It must direct the light of ordinary intellectual awareness on the inner world of the life force and universal consciousness. It must illumine the illuminator, and the "observer" in scientific research. It must reveal the connection among all things, showing emphatically that our place is everywhere. Such a vision will give new meaning to seventeenth-century John Donne's, "No man is an *Island*, entire of itself." The required new "myth" will bring about a reunion between the scientist and the mystic, the rationalist and the intuitive, and the followers of narrow tenets and those with no tenets at all.

Joseph Campbell became the 1980s guru of a movement to rediscover our mythological past and interpret it in terms of twentieth-century experience. In a PBS interview with Bill Moyers, he stated, "myths are clues to the spiritual potentialities of the human life." Moyers rejoined, "myths are stories of our search through the ages for truth, for meaning, for significance." Both were correct, but if the myths of the past do not accommodate new knowledge about ourselves and the natural world, they are no longer credible.

In this context, myths are the stories we tell ourselves to give coherence and meaning to the experiences of life. They are composed of combinations of fact, fiction, and fictionalized fact, but they make sense to those for whom the stories are told. Such myths can be used to describe historical events in a manner that serves to explain certain beliefs, practices, and institutions in a society, as in the myth of Moses and the Children of Israel in Egypt and their escape across the Red Sea. A myth may embody a more fundamental idea, such as a Native American creation myth, or a set of cultural values, such as the mid-western myth of Paul Bunyan. A myth may include a veiled explanation of an enduring truth, as in Plato's allegory of the cave or Jesus's parable of the mustard seed.

Humanity needs a new "humans-in-the-cosmos myth," a new description of our Solarian Legacy, one that is powerful enough to encompass believable explanations of our stellar past, includes theories for things unseen, and leads to significant behavioral choices. It will deal with such profound issues as the creation of the universe and its creatures, evolution of social institutions, and human relationships with nature and other beings. Its recitation will connect seemingly disconnected events to reveal their inner meanings. Such a myth for the twenty-first century will recall the perennial wisdom of the past, consolidate and integrate the lessons of many cultures, and take advantage of the insights of science that can be validated through human experience.

In the anomie of the current age, a limited renaissance in consciousness has already begun. Alternatives to fundamentalism have increased in popularity, even as fundamentalism's adherents increase among those who are most threatened by change. Growing numbers turn to Eastern mysticism. Traditional beliefs in goddesses and shamans now attract large groups.

Native American teachings and esoteric mystery schools (Egyptian, Greek, Muslim, Jewish, and Christian) are being offered to broader publics. "New age" beliefs are found attractive by others. But, as with the deficiencies of fundamentalism, the limited and divisive nature of many of these belief systems is being countered by the research of serious frontier scientists and nondogmatic metaphysicians.

Even with numerous constructive cultural and historical insights published by independent and innovative writers, some of whom are highlighted in this book, there is no new "humans-in-the-cosmos myth" emerging that can draw all societies together. In the 1990s no intellectual or religious vision has provided a framework to unite the human race around a few core understandings that would facilitate global cooperation on the issues of peace, health, social progress, and ecological well being. *Metascience*, incorporating the best available knowledge from frontier science and traditional wisdom, can form the basis for such a consensus, creating *a new cosmic myth that resonates with the inner experience of most and becomes a powerful guide for the behavior of all.* As parts of this composite vision ring true, people will be more inspired to test the rest of it.

Although the definitive material or intellectual consensus on which to build such a new cosmic myth is not yet available, abundant evidence from new and ancient sources does exist to expand our frame of awareness and help us see more clearly our part in the stellar or galactic story. *This book presents solid research underpinnings for a provisional new story,* and offers the rudiments of the myth that can accelerate the advent of the New Renaissance.

PART I

Three Perspectives on the Universe

MODERN SCIENCE THROUGH THE POPULAR MEDIA has supplied us with information and images of the vast and strange cosmos we inhabit as Solarians. Part I focuses this ephemeral vastness in the mind's eye from three perspectives: the largest scale possible, the physical universe in all its macrocosmic glory; the microcosm beyond quantum physics in the most infinitesimal terms imaginable; and the human home base Earth, an interdependent member of its stellar and galactic neighborhood. Part I identifies the need to expand the conventional scientific perspective.

Chapter 1

Macrocosm: Conscious Universe

Physicists and astronomers have researched backward in time to uncover the material history of an explosive event billions of years ago that resulted in today's universe. It is a marvelous portrayal of unimaginable forces swirling bits of matter across light years that slowly settled into stars and planets, eventually producing human beings. But the story has large gaps: From where did the original egg come? How did the principles that govern such complexity find their expression in inert matter? When and how did consciousness enter the picture? Conventional science doesn't have the answers. Intelligent use of metaphysics and alternative ways of knowing can help fill in the puzzle.

◆ ◆ ◆

MY FIRST MEMORIES AS A CHILD were of the macrocosm, although of course I didn't know the term. The sunlight streamed through the cheap cotton curtains to warm me as I lay on the hand-sewn quilt Mother had made for me. Dogs barked at the occasional wagon or truck that passed on the rural road. Birds noisily ate figs from the tree outside the window. Wind sang in the pine trees overshadowing the house. The sky darkened as clouds moved across plowed fields. I saw lightning and heard thunder just before rain began to pelt the tin roof. I felt the spirits lurking around us and the souls of ancestors lounging higher up. As a small country boy, I experienced this seamless reality in which animals, sky, earth, heaven and hell, family and ancestors were all integral to the universe created by God for his purposes.

Growing up in the 1940s in a poor, backwoods section of Northwest Florida, I felt the cycles of my life as parts of this larger whole, sensed myself inextricably linked to the daily rising and setting of the sun. It lit the morning sky even before the blazing ball itself appeared on the horizon, marking the hour to feed the animals. Its warmth thawed the ground in preparation for planting as its movement north made the days grow longer. The waxing and waning moon determined the planting schedule; its magnetism affected the response of seeds to the earth just as it affected the fertility of the women and the female animals. Sitting on the porch after supper, we would start our night watch with Venus in the evening sky, anticipating the calls of the whippoorwill and the hoot owl as we talked of all the beings touched by the same god. We could sense their presence, just as we felt the breezes evaporating the sweat remaining from last-minute chores. As the Big Dipper and Orion's Belt became discernible in the darkening sky, we were confident of our understanding of it all.

Seamless Cosmos

Our rustic perspective was not significantly different from that of shepherds on pre-Christian Middle Eastern slopes, or Australian Aboriginals on a walkabout following their "dreaming tracks" in the outback, or Native Americans planting and harvesting with seasonal rites keyed to the movement of the constellations. Each was compatible with the ancient Egyptian song of the earth as giver of all gifts, itself blessed by the sun that quickened the rivers and flowered the deserts. For all of us, the living earth was subject to the living sky; we knew we were children of the heavens. Nothing was dead, nothing was separate. Our lives, except for a limited ability to maneuver among daily events, were shaped by forces beyond our control.

This basically naive but comprehensive view of the interactive nature of the cosmos* has for millennia dominated the perceptions of traditional

* Throughout the book the terms "cosmos" or "cosmic" implies something larger or beyond the material universe as we know it.

peoples in the world. However, since the time of Aristotle (*On the Heavens* written in 340 B.C.)—the Greek philosopher from whom is dated so many intellectual concepts—a more restricted view has become dominant among so-called developed peoples. A stream of ideas spreading from ancient Greece westward to Rome and up through Europe and over to the Americas has shed light on many parts, but has reduced our understanding of the whole. Propagated by the Anglo-Saxon and Latin-centered world of thought and technology, Western civilization's mechanistic view of the universe has been a two-edged sword: As scientists dissected the universe, they excised sectors of human experience from their scrutiny.

Aristotle's theory of solid spheres containing various heavenly bodies rotating in fixed circles around a stationary earth was followed by the discoveries of Ptolemy, Copernicus, and Galileo. Galileo's observations in 1609 with his newly developed telescope indicated that not all heavenly bodies were orbiting the earth, or the sun. People began to perceive that moons orbited planets that in turn orbited the sun, and that suns and stars had their own tracks within galaxies. Isaac Newton's theory of gravity, published three-quarters of a century after Galileo's telescopic observations, provided an explanation for the spinning, elliptical movements of heavenly bodies.*

As telescopes became more powerful, people saw more stars. But they continued to assume they were looking at a largely static universe, with the human observer at its focal point, set in place by a supernatural being. No one knew how the world and its universe got started. Aristotle himself postulated the theory of an undefined "Prime Mover." The religions of the West—Judaism, Islam, and Christianity—believed they were started by the divine creative act of a personal god.

Between the seventeenth and twentieth centuries, many philosophical arguments appeared about the nature of creation, the limits of the universe, and the issue of time. Yet the basic perception of the macrocosm remained the same, whether people believed that it had always existed or that it had been created fully blown at the beginning of historical time. From both these perspectives the universe was something whose laws could be discovered and whose elements could be manipulated. Whether the laws were

---

* Johannes Kepler realized the orbits were elliptical, not perfect circles, in the early seventeenth century.

mechanical or divine, they were all fixed and susceptible to understanding by rational inquiry.

This perception of a material universe with fixed boundaries was shattered in 1929, when Edwin Hubble (after whom the orbiting Hubble Space Telescope is named) saw that other parts of the universe were moving rapidly away from us. (Five years earlier he had discovered galaxies beyond ours.) Hubble interpreted such movement to mean the universe was expanding, and if it were expanding, it had to have a history of considerable duration and speed. These conditions could result from either a single creative event such as the Big Bang, or a continuing process of external influence. Since the latter opens difficult questions about the nature of unknown forces outside the universe, most scientists have chosen the safe route and accepted the Big Bang theory. As a result, humans are still locked into very limited assumptions about themselves and the inner nature of cosmic reality.

Scientists continue to fragment the search for knowledge, separating it into various disciplines and categorizing some human experiences as natural and normal and dismissing others as supernatural and paranormal. Most aspects of astrology are ignored by mainstream science, although the experiences of untold millions indicate strong correlations between human behavior and the actions of celestial bodies. Likewise, the link between thought and the microcosmic activity of cells is largely ignored, as is the whole phenomenon of telecommunication among conscious beings. Science's focus on selected "natural laws" comes at the expense of others, particularly those dealing with the human being—both as subject and object.

Some steps toward synthesis are being taken, however, by forward-thinking professionals in physical science, archaeology, anthropology, psychology, and consciousness research. They are joined, despite institutional barriers, by nonsectarian mystics to expand the timeframe of assumptions about human history and definitions of matter and consciousness. Intuitively each is rediscovering the singular universe of traditional peoples, the seamless reality of childhood.

The collective challenge in enhancing overall understanding of the origins of the universe is to glean from tradition any insights that mesh with the picture emerging from frontier science. Some universal traditions may be helpful. The Judeo-Christian tradition proclaims "in the beginning was the Word."* The Australian Aboriginals believe the ancients sang the world into being.[1] The concepts are the same: From the word comes the thing. Creation and development follow ideas. New ideas lead to new creations and changes in the course of development.

Most mythic traditions—whether from Central America, North America, India, China, Egypt, Greece, or the Middle East—include allusions to a being or force that formed something out of nothing. The truth is that we do not know how it all started.** What we do know is that the creative force that gave birth to our universe is more powerful than the energy of the suns and stars of a thousand billion galaxies and more intelligent than we can fathom.

*In that incomprehensible birth moment, that fraction of a "first second," exists the potential for all that we know and all that is to be uncovered in the eons to come. This potential comes in a stew of subatomic particles and waves of energy, to realize itself in polar opposition: matter-energy and its bonding forces juxtaposed with antimatter-energy and its repelling forces.**** These forces are part of something greater than and beyond the materials from which our universe was born.

Principles embedded in the cosmic egg, laws that will continue to affect all the creation to follow, partially manifest themselves in the very first trillionth of a second of ordinary time. What are they? Modern scientists have been trying to discover these principles one by one. Earlier civilizations have developed their own systems of explanations. One such system, the Hermetic Principles, has come down to us from prehistory.

---

* Out of that word, with "no place," "no time," and "no thing," came the birth of our universe. The New Technology Telescope (NTT) in La Silla, Chile, has seen backward in time through what many assume to be a universe that is 20 billion Earth years old. The repaired Hubble telescope, orbiting the Earth, will extend and refine this reach.

** See David Darling's poetic story of the beginning of material time.[2] Most current estimates of the age of the universe range from 15 to 20 billion years.

*** In this chapter the parts that reflect advanced, but conventional, scientific views of the history of the universe are in italics. Hermetic Principles relevant to the contiguous text are shown in parentheses.

Hermetic Principles

The Hermetic Principles, named after a legendary personality, Hermes Trismegistus (meaning "thrice great"), have been known and articulated in esoteric scientific circles for more than 5,000 years. For many centuries the term "hermeticism" was inappropriately associated only with alchemy, or the alleged transmutation of metals into gold. By focusing purely on the material realm, such references obscured the deeper mental, energetic, and spiritual meanings of this ancient wisdom. Today the term "hermetic" has come to mean secret or sealed.[3]

Hermes Trismegistus, identified by some as the god Thoth, was one of the "wise beings" who shared knowledge and insight with the pre-pharaonic Egyptians. Hermes is considered by many to be the source of the basic teachings that infused all the highly intellectual, esoteric traditions of the Egyptians, Greeks, Jews, Muslims, Hindus, and Christians.* A few learned initiates guarded the insights** and passed them on discerningly over the centuries to those deemed ready for the teachings. During the Inquisition and other periods of religious persecution, it was dangerous to reveal one's belief in an alternative reality. Consequently, most of the understanding was lost on the masses, as well as the scholars and students of the modern era.

The seven Hermetic Principles are Mentalism, Correspondence, Vibration, Polarity, Rhythm, Cause and Effect, and Gender.*** (See Table I.) Although they can appear to be so simple and mundane that the casual reader is wont to skim them lightly, they may actually be more far-reaching than the basic assumptions of Newtonian mechanics or quantum physics. The current work of various frontier scientists and grounded metaphysicians is confirming the validity of these concepts. We can now begin to test them in an integrated way to determine their application to all dimensions of our universe.

---

* Some believe the same or similar being was a source of knowledge for the Toltecs, Mayans, and Incas of the Western Hemisphere. For example, the Mayan calendrical system had several principles similar to the Hermetic ones presented here.

** This tradition is the basis for the plot of the currently popular *The Celestine Prophecy*.

*** In Hindu metaphysics we find similar principles: The primordial sound of Aum indicates the Principle of Vibration, while the concept of Brahman coincides with the Principle of Mentalism.

Table I

Hermetic Principles

	Principle	Description
1.	Mentalism:	Everything exists first as an idea.
2.	Polarity:	There are two aspects to every phenomenon
3.	Correspondence:	The same fundamental rules apply at all levels.
4.	Vibration:	All elements of the cosmos are in constant motion.
5.	Rhythm:	Each entity, energy, and idea has its own cycles/patterns.
6.	Gender:	Yin and Yang, receptivity and expressiveness, exist at all levels.
7.	Cause and Effect:	All aspects of the cosmos are in a singular interactive system.

Mentalism

The Principle of Mentalism is reflected in the Biblical quotation, "In the beginning was the Word." Ultimately all external reality is based on idea or concept. In the context of quantum physics, Mentalism means the physical world can be reduced to patterns of potential connections among potential concentrations of matter/energy that might or might not come into form, depending upon the introduction of consciousness.[4] Twentieth-century physicists and consciousness researchers are thus on the edge of unraveling the implications of Mentalism that Hermetic initiates have known all along. Now anyone can grasp its meaning: the basic force in the universe is mental.

Correspondence

The Principle of Correspondence, "as above, so below," means that one can infer the nature of distant realms from local experience. The dynamics of cells are parallel to the dynamics of galaxies. Just as a small laboratory or computer program can simulate the behavior of stars billions of light years away, the consciousness of an individual being can confer with the Ultimate Consciousness that existed when there was only the Word. This principle ensures, for example, that humans need not dread exposure to the ideas of, say, extraterrestrials; they are derived from the same universal consciousness.

Vibration

The Principle of Vibration, which asserts that everything is in continual motion, is now a basic tenet of science. Subatomic particles are continually moving in relation to each other in every concentration of energy and mass in the universe. The patterns of vibration occur in all manifestations—from dense stone, to gaseous molecules, to the thoughts and emotions of human beings. We have intuitively grasped the validity of this principle: we get "good vibes" about this or that. When we are on different frequencies with someone, we can wind down or increase the tension, thereby moving a situation to a congruent level of vibration.

Polarity

The Principle of Polarity embodies the truth that two seeming opposites are in truth complements that differ only in degree—the obverse and reverse sides of the same coin. This principle applies in all realms. Photon particles are inextricably linked in pairs, with each as either the positive or negative aspect of the other. Hot and cold are but different aspects of the same temperature gradient. Any characteristic in nature or cosmic experience has its own gradient—large and small, high and low, black and white, sharp and dull, male and female. Where does each pole end and the other begin? What about the shades of good and evil? The crucial point here is that all such polarities are only different vibrations on the same continuum. One can be transmuted into the other employing the Principle of Polarity.

Rhythm

The Principle of Rhythm means that everything manifests itself in a pattern of to and fro, up and down, in and out. The movement in one direction is always compensated for by a return. For every action there is a reaction and for every advance there is a retreat. The principle applies in all the affairs of the cosmos—stars, beings, mind, energy, and matter. It works in the interactions within a plane, and in communications between dimensions. Over time, the rhythms result in spiraling shapes that characterize much of the universe.

Understanding of the dynamics of this principle makes it possible to mitigate some of its more extreme effects. We can recognize that fatigue, followed by rest, leads to renewed energy. Anger gives way to remorse and pain succumbs to release. By being aware of the rhythms, one is less likely to resist their flow, thereby reducing the buildup of extremes.

Cause and Effect

The Principle of Cause and Effect is more commonly known by its ordinary meaning: *x* acts on *y* and causes *z*. From the Hermetic perspective, to say each effect has many causes is more accurate. This multilevel reality is epitomized by Carl Jung's use of the word *synchronicity* to describe events that, though outwardly appearing to occur by chance, are actually *the inner workings of one or more cosmic laws.* Indeed, all events are at some level the workings of cosmic law. What we attribute to "chance" is usually an event whose governing law is not self-evident. True chance or randomness probably occurs only at the level of quantum gaps, where there is a true break between past and future.

The Hindu concept of karma is an illustration of the principle of Cause and Effect, as is the Christian admonition "as you sow, so shall you reap." Human societies are only now learning the dramatic effect of this principle in ecological systems. Now humankind must become more aware of cosmic law in the realm of consciousness, in its role as Conscious Co-Creators of the Universe.

Gender

Gender, the last Hermetic principle, has remained most obscure because we tend to equate gender with primary physical sex characteristics.* However, every being and every plane in the cosmos contain the dual elements of Yin and Yang, feminine and masculine. The term *gender* recognizes this "complementariness" within all self-contained units of the universe. Even in apparent single-sexed entities, one aspect is the receptive nurturer, while another is the expressing creator. The Principle of Gender itself obeys the Principles of Polarity and Rhythm, in one circumstance manifesting the masculine aspect and in another the feminine. Neither is ever totally absent: in space-time balance is assured. Fully aware cosmic beings seek harmony in living their dual nature (Gender), honoring the ebb and flow (Rhythm) called for by the organic developments in self, society, solar system, and cosmos.

These seven principles are simple keys to the mysteries of matter-energy, spirit-mind, and consciousness. They can open the gateways through which a profound transformation of conscious life becomes possible.** This book is an argument for undertaking such a journey, demonstrating that transformation on the mental and energetic planes will have immediate consequences in the material realm. Not one principle stands alone: all affect each other in a mode of reciprocation, thereby assuring the cohesion and unity of the multifaceted universe. (See Table II for applications.)

---

* See Ivan Illich's book *Gender* for an excellent portrayal of the distinction between gender and sexual characteristics.

** The experience of jazz, by both the musician and the listener, is a mundane illustration of this interplay. Vibration and Rhythm are communicated through sound and sight. Gender is evident in the artistic expression and the admiring audience. The continual creative act of Mentalism finds its way into the Polarity of sound and silence where the Correspondence of scales manifests in several instruments. Music that is universal taps into the vibrational signature of a species and gives rise to emotions, health, communications, and a sense of community.

Table II

Application of Principles

Hermetic Principle	Physical Events
1. Mentalism:	Superstrings Periodic Table
2. Polarity:	Positive/Negative Charges Matter/Antimatter
3. Correspondence:	Atoms and Star Systems Cells and Families
4. Vibration:	Electromagnetic Spectrum Other Spectra
5. Rhythm:	Birth, Life, Death Creation, Elaboration, Decline
6. Gender:	Male/Female Expressing/Sensing
7. Cause and Effect:	Warmth to Sprouting Love to Creativity

Everything Pulsates

A return to that first burst of ordinary matter will show how the Hermetic Principles function both sequentially and simultaneously in the evolution of the physical universe.

Scientists feel confident that somewhere between fifteen and twenty billion years ago there was an explosion (Vibration) *into form, a Big Bang, when out of "no place," "no time," and "no thing" came the birth of our universe. At this moment, from an egg-like but invisible point, came streams of protons, electrons, and neutrons.*

The electron is almost nothing; the protons and neutrons are 1800 times heavier. But there are many electrons swirling around, vibrating in emerging space. Each electron has an antiparticle called a positron (Polarity).

Each electron or positron is equal in mass to its respective twin, but the electric charge each carries is reversed. The positron's charge is just as positive as the electron's is negative.

They move about forming the stuff of ordinary matter, entwined in the cosmic dance of creation, appearing to be opposites (Gender), *but in reality only two halves of a pair. As long as they keep the right distance, held in position by the opposing electric charges* (attraction of opposites or Polarity), *they function as matter. When they clash, they destroy each other.*

In destroying each other, electrons and positrons create a pair of different particles—new photons. Thus death leads to birth: that which disappears returns in another form, manifesting the Principle of Rhythm.* These new photons are different from the old electrons; they are particles of light, which is pure energy. (And, as will be discussed later, light can act as either a particle or a wave.) Given this dynamic, why didn't all the electrons go up in light billions of years ago? The rhythmic principle insures an overall equilibrium: when two photons collide they give birth to two new particles of matter.

Parenthetically, it is worth noting that Genesis 1:1-3 has the order of creation correct. First was chaotic matter. Then "God said, 'Let there be light,' and there was light." Photons followed electrons, neutrons, and protons. The creation and destruction of subatomic particles according to Genesis involves a repatterning of pure energy, thus illustrating the Principle of Mentalism. In other words, the *logos* or idea expressed by the conscious creative force shaped its world.

Theoretical physicists believe they can calculate back to the point when the universe was 10^{-43} *seconds old.* 10^{-4} *is one ten-thousandth of a second on an exponential scale. So* 10^{-43} *is only one hundred million, trillion, trillion trillionths of a second. At this "Planck time" (named after a founder of quantum mechanics) all forces and matter acted as one unified force. Because mathematical calculations do not work beyond that point, physicists assume that from that point gravity split off from the singular force*** *that*

---

* The concept of reincarnation of souls (local manifestations of consciousness) mirrors the same principle on a higher level of complexity.

** The four forces dealt with in physics are discussed more fully in Chapter 4.

hypothetically existed in the beginning (Principle of Cause and Effect facilitates inferences).

At the 10^{-35} *second (getting older), there was only gravity and another apparent force that combined the currently understood electromagnetic force and the weak and strong forces* (Gender). *The universe was pure energy, with point-like particles of quarks and leptons. (See Chapter 3.) Matter and antimatter were equally balanced* (Polarity). *Physicist Blas Cabrera at Stanford University hypothesizes the existence of monopoles or free magnetic poles that formed another kind of matter* (Gender).

At 10^{-32} *second the universe was only about the size of a grapefruit; gravity and the strong force now stand with the electro/weak force (electromagnetic and weak forces still combined). At* 10^{-20}, *black holes may have formed (Polarity). At* 10^{-12} *second the temperature of the universe was 1,000 trillion degrees. At* 10^{-10} *second, according to a "hot" bang theory, the universe was about the size of our own solar system. At this point the four forces labeled by modern physics were distinguishable from each other.* (The Principle of Gender comes more fully into play.) *Between* 10^{-6} *and* 10^{-4} *the stew of quarks began to fall into triads and form neutrons and protons, elementary particles that coexisted with leptons.*

At 100th of a second the universe had cooled to 200 billion degrees Celsius. Hundreds of types of particles were engaged in the cycle of birth and death and re-creation (the Principles of Rhythm and Gender).

At the end of one second, the universe was a bubble of space only 200,000 miles across according to a "cool" bang theory. Its temperature was 10 billion °C. The antiprotons and the antineutrons have gone. On another track, some electrons have merged with protons to yield neutrons and neutrinos. The latter are so infinitesimal in mass, if they have mass at all, that they can achieve almost the speed of light and can pass through the most dense matter unimpeded.

The Hermetic Principles pose intriguing questions. What if the antiparticles formed the black holes scattered around the universe?[5] Do current assumptions adequately take into account the role of antimatter? Is a new cycle of creation (Rhythm) started when antiparticles rejoin and destroy their twins in the world of ordinary matter? Is that how other universes are formed?

Stephen Hawking[6] has attempted to better understand the dark side (Polarity). He has postulated the existence of billions of tiny black holes (necessarily formed in the early fractional second when pressure and temperature were high enough). This theory marked a shift from his and others' earlier view that black holes were caused only by collapsing stars. Hawking has broken with another premature assumption and now believes such holes may emit energy and may explode. These attributes are predictable from the Hermetic Principles of Polarity and Rhythm.

During these early nanoseconds (one billionth of a second) of the universe, time as we know it did not exist. With such concentrations of mass and energy, developments occurred at an exponential rate. As much could happen in the first one tenth of a second as happened in the second, and the first ten seconds, and then the first 100 seconds, and so on (a function of the scaling up of the Principle of Correspondence).

The initial explosion produced unimaginable heat, but as the seconds turned into minutes, things slowed down and cooled off. The subatomic particles coalesced into elements—ordinary hydrogen, heavy hydrogen (deuterium), and the heaviest hydrogen (tritium). Next, the various densities of helium came into being. On and on through the periodic chart, elements formed as particles bonded according to some a priori *set of inherent principles* (Mentalism), *indicating at this early stage the impact of conscious order.*

Three to four minutes after its birth, the universe was filled with radiation, caused by electrons destroying almost all the positrons (Cause and Effect). *The strong force started forming the nuclei of the above-mentioned heavy atoms. At about this time, a hydrogen/helium ratio of 3:1 is believed to have developed* (Rhythm).

At the end of 30 minutes, the temperature was 300 million degrees (only 15 times hotter than our present sun). The average density of everything was less than one tenth of water.

As the hours turned to days and years, space expanded, but the quantity of matter stayed approximately the same. For thousands of years there was only a mist, charged and swirling in a huge electromagnetic field.

Patterns Emerge

While the collective theories of modern physicists, italicized in this chapter, may provide useful descriptions of the universe's development, they cannot be considered the whole story. The Principle of Mentalism implies something other than chance was at work in the mist. Scientists have not hypothesized an *a priori* design inherent in the speck of original sperm and egg from which the rest of phenomenal matter burst forth. Had humans been around, they could have seen the patterns emerge, but now we can only attempt to reconstruct them from the traces that remain. Research in physics has explicated some of those patterns, but, as this book demonstrates, important parts of phenomenal reality (and most of nonphenomenal reality) are still unexplained.

Physicists and astronomers in NASA's COBE program (Cosmic Background Explorer) have measured the apparently ubiquitous background radiation (long wavelength microwaves) spread throughout the universe. In April 1992, they reported a team had measured minuscule temperature differences (one hundred thousandth of a degree) in this radiation. The difference is analogous to ripples on an otherwise smooth pond surface, but in this case they are ripples in the fabric of space-time. Thus, more than 15 billion years ago, shapes (termed "fossils" of creation by astrophysicist George Smoot) began to differentiate themselves in the primeval fog of the universe. What caused those ripples is still unknown.

Even with great scientists, mindsets sometimes get in the way of increasing knowledge. Einstein believed so fervently that the universe was stable—not expanding or collapsing—that he initially adjusted his own perfectly working equations for his theory of general relativity in order to support his bias. Stephen Hawking, the highly-reputed contemporary theoretical physicist, dismisses solid evidence from parapsychology—misnamed because the phenomena studied are validated by human experience. Such a deliberate exclusion of the role of consciousness in creation calls into question the unfolding of the Big Bang theory and related hypotheses. The recognition of the influence of mind on subatomic particles, at the core of quantum physics, is opening the door to a bigger role for consciousness in theory building.

A complex universe could not have randomly evolved from the protomaterial and nonmaterial elements were they merely floating around in absolute chaos. Certain pre-existing, unseen forces or dynamics were essential in bringing form from the primordial stew (Mentalism; Cause and Effect). Modern scientists have discovered four such forces—gravity, weak nuclear force, electromagnetism, and strong nuclear force—which they are now attempting to reduce to a single principle called GUT, or Grand Unifying Theory. But why assume that all operative forces have been discovered? Is science once more falling into the old Cartesian trap of dividing mind from matter, and then asserting that matter created mind? While matter is coextensive with mind (having the same scope and duration) and cannot exist without it, the mind-over-matter postulate (Principle of Mentalism) can be validated in human experience and in quantum physics.

Scientists, in the rush to reduce their principles to one GUT, conclude that all forces except gravity may be a single force operating within the atom. In 1979, three scientists* received the Nobel Prize for experimentally demonstrating that electromagnetism and the weak nuclear force (that which controls radioactive decay) were two aspects of the same force. Subsequently, others have tried to demonstrate that the strong nuclear force (that which holds the nucleus of atoms together) is another dimension of that force.

To the extent that these conclusions are not congruent with the Principles of Polarity and Correspondence, they will likely prove to be only partial explanations for observed phenomena. (See Chapter 4 for further analysis of the gaps not covered by the four forces.) Since the four forces do not explain all observed experience, nor all the seven Hermetic Principles, we must continue our analysis of the developing universe before we presume to have an all-encompassing theory.

Within the seething mist, differentiation began to occur; localities seemed to assert themselves. While much of it continued to expand in all directions, small areas began to coalesce. Something seemed to be luring bits of matter together. Only within certain frequencies could the electro/weak/strong force triad come into play. What was controlling the frequency shifts? Were they a

---

* Steven Weinberg, Sheldon Glashow, and Abdus Salam.

function of random temperature changes? Were they the result of "grand" patterns implanted in the embryonic universe before its birth?

While we do not know the origins of certain inherent patterns, we do know that throughout this mist of ordinary matter they could be discerned through the manifestation of shapes. The Hermetic Principles of Vibration and Rhythm seemed to be operating as the frequencies of ordinary matter changed (most slowing down) and masses became more dense. In these early stages, the rhythmic pattern involved exceedingly long time frames.

Some scientists say the coalescing was a function of the strong force splitting away from the remaining electro/weak pair. They postulate the existence of cosmic strings—great loops of tension composed of energy moving at near-light speed—that broke up the uniform mists. Initially small strings combined in ever-increasing sizes until they became powerful enough to draw large collections of particles together.

Such a descriptive theory for the early condensation of matter still falls short of an explanatory theory: At what point and how were the cosmic string patterns ingrained at the event of the great birth, or rebirth? As the strings became larger and longer and put wrinkles in the smooth surface of space-time, they gradually disappeared. None exist for us to see today at the cosmic level, but their analog—the DNA helix now so familiar to us—still works its way (Principle of Correspondence) in the "creation" of plant and animal life.

Thirty thousand years after its birth, the infant universe was composed of matter and light clouds kept intact with light energy.

At 300,000 years (keep in mind the exponential time line), the universe was bigger and cooler, the temperature dropping below 8,000 degrees and the color moving from hot colors toward the green and blue side of the electromagnetic spectrum.

At 500,000 years, electrons and nuclei from the primordial stew began to form permanent attachments within the shapes of atoms. As matter coalesced it increased in complexity, including the separation of dark matter from ordinary matter.

Opposites Attract

What is this division of matter into two qualities or properties? The dark cannot be seen, but it exerts a powerful influence on the behavior of the universe. Is this the Principle of Gender at work, with its receptive/nurturer aspect juxtaposed with the expressive/creator aspect? The black-and-white yin and yang symbol aptly illustrates this dynamic interaction.

While one aspect holds sway in a particular region of space-time, neither is totally absent. Overall balance is maintained in the universe. Is the Principle of Polarity also at work, insuring that effects in one area (birth or death) are absorbed or matched by the opposite pole in a contiguous part of the universe?

One hundred million years into its existence, the universe begins to seed protogalaxies. Patterns of connections appear as the unseen dynamic that weaves matter and antimatter, strings and mist, mass and energy, light and darkness into the components of a vibrant universe. From a protogalaxy will come one galaxy—to be called the Milky Way—which is centered in the direction of a constellation to become known as Sagittarius.

But we're getting ahead of the story. The Principle of Correspondence ensures the similarity of patterns in atoms and planets, in molecular and star systems, and later in various species of conscious beings. New stars pop into view, glowing in a field of thinning dust, but they are not alone. A force or guardian is always nearby.*

---

* Science is beginning to grasp the nature of this balance on a cosmic scale. In April 1992, NASA reported evidence of many new black holes. They are no longer viewed as an isolated phenomenon created only where an old star collapses on itself. They cannot disappear, for they cannot become smaller than the original

Many observers now believe each galaxy, including our own, has at its center some form of black hole. Black holes both suck up and expel matter, and the power of this dynamic may play a role in the rhythm of time. Some speculate that black holes are interdimensional channels through which time travel could occur. General relativity theory says mass affects time and space. In a gravitational field of this scale, space and time behave like matter. So black holes with such mass could pull light toward itself, speeding the light up to fade into the future. This possibility is plausible if the so-termed "worm hole"—a black hole linked to a white hole, its Hermetic polar opposite—expels in the reverse plane all that it swallows from this plane. (A few scientists speculate that such consumption reduces by half all the knowledge accumulated in the universe. When the matter is regurgitated it is without memory, ready for a new cycle of learning.)

The phenomena of black and white holes are cosmic-scale demonstrations of the Principles of Polarity and Gender. Emissions in infrared wavelengths through the galactic dust indicate great bursts of energy as matter is drawn into the maw of a black hole. A black hole is so powerful that it can swallow a star that comes too close and hold in the light, resulting in high winds that create a million-degree black bubble.

Now, a billion years after the Grand Rebirth (explained in the next section), different types of galaxies are evident: A few appear to have quasars at their center, many have ordinary stars, and some have black holes. Others have spirals at their core. All are different, but clearly of the same species.*

Our galaxy, the Milky Way, is approximately one hundred thousand light years across, only one of one hundred thousand million galaxies—more or less—each in turn containing about one hundred thousand million stars. All the other galaxies appear to be rapidly running from us. We believe

---

primordial universe. Some are quiescent. They come in all sizes, like light bulbs. One, 3 million times the mass of the sun, may be only 2.3 million light years from the earth. Some now believe certain black holes could be spread out so thinly that we could pass through them without knowing it.

* Quasars (from quasi-stellar) are bright, starlike entities that convert matter to energy and emit gamma rays, the highest frequency on the electromagnetic spectrum. The quasar's energy output could be a thousand times stronger than that of our entire galaxy and ten trillion times more potent than our sun.

this because, in a manifestation of the Principle of Vibration,* their colors become more red (as the wavelengths become longer).

After five billion years, many of the stars in most galaxies have burned up all matter available around them. Some of them become so clogged with heavy matter that they collapse and then explode into supernovas (Rhythm). *This phenomenon results in the creation of dozens of new elements, which in turn become the building blocks of evermore complex life forms.*

The black holes, quasars, and other less powerful poles apparently reflect the dynamic tension of balance suggested by the Principle of Polarity. The Principle of Rhythm gives pattern to the play of these galactic poles over the eons.

Another ten billion years pass. More stars are born; many die. Much matter is pulled together into the various shapes of galaxies, but intergalactic space is still filled with a flurry of tiny specks—the material from which worlds are created.

The Hermetic Principle of Gender is at work from the cellular to the galactic level. In the latter, a dark fertile (feminine) force dances with a light mercurial (masculine) force in the rhythm of destruction and creation. If the appropriate distance is maintained, the possibility of offspring exists. Due to the Principle of Cause and Effect, we can see the products of this cosmic dance of stars and dark holes: their children are born as new planetary systems.

Although other planets may have formed earlier, our Earth was born about five billion years ago to an average-sized yellow star on the fringes (28,000 light years from the center) of the Milky Way. At the edge of that galaxy, in a backwater eddy of this gigantic maelstrom, cosmic forces drew scores of stars larger than our sun to the brink of a black hole. These same cosmic forces spawned a chain of events that eventually gave birth to conscious beings.

* The modern term used to describe the principle is the Doppler Effect: as the source of light moves away from us its frequencies appear to slow down, like the sound of a horn moving away in the distance.

Back to Zero

While most scientists accept the idea of an explosive birth for the universe, heated debate still continues as to whether the universe will perpetually expand or, some eons from now, will reverse direction and collapse back on itself. It is assumed to be currently expanding about 5 to 10 percent every billion years. Given what we know of forces at work and the amount of matter (star systems and dark holes), some scientists postulate that the universe will continue to expand forever.

The Big Bang theory is supported by some evidence of a uniform presence of background microwave radiation throughout the universe (although that is incompatible with some current assumptions of quantum mechanics and recently discovered evidence pointing to minute variables in the radiation). The 3:1 ratio of hydrogen and helium, assumed to have been developed in the first few minutes of the universe, can be observed as far as we can measure. The universe looks the same in all directions, even though we are not in the center of it (Principle of Correspondence).

Some scientists foresee a flat, uniform universe, while others foresee a subsequent contraction or a Big Crunch. Whether the expansion is indefinite or time-bound seems to depend on there being a certain threshold of density to matter. Many cosmologists believe the universe to have exceeded that critical mass and therefore believe it will cease expanding. The dynamics of galaxies predicted in Einstein's 1916 theory of relativity (great mass can exert enough gravitational pull to bend light waves) may indicate the presence of unseen or "dark matter" among visible galaxies. For example, the light distortions seen by the Hubble Space Telescope in group NGC 2300 at 150 million light years from earth, indicating a gravitational force 30 times that which we experience, add credence to the contraction view. This kind of "dark matter" may keep the visible matter from expanding indefinitely.

Jonathan J. Halliwell identifies some of the problems with the Big Bang model of the universe.[7] For example, a number of vast regions are moving away from each other at a rate that appears to indicate they could never have been in contact with other parts of the universe in their entire history. But why are these unconnected parts so similar? Why did the universe stay so flat if there was a uniform explosion from a single point? How did small-scale fluctuations in the near-zero moment lead to the later large and varying structures? In this situation, an explicit assumption of predesigned "cause and effect" relationships makes more sense than belief in a random explosion. The Big Bang theory does not offer an explanation for either the source of the single point that exploded or the origin of the space into which it exploded.

A new scientific theory posits that, at a fraction of a second after point zero, something (called "inflation" by physicists) acted on the expanding explosion that set patterns in place that resulted in the phenomena (star systems, planets, and human beings) we observe today. The "superstring" theory holds that infinitesimally small loops, 100 billion times smaller than a proton, could lead to such complex organisms evolving in the undifferentiated space. Once again, it stretches credulity to make the leap from a "random bang," through chance "strings," to "complex conscious beings." This is the corner physicists have painted themselves into. One escape for them is a theoretical combining of general relativity (Einstein) and "singular theorems" from quantum assumptions (Hawking and Penrose, respectively) that makes it possible to postulate a point close to the universe's birth when "external" influences could have impinged on the universe's development trajectory.

In effect, these scientists are now recognizing the feasibility of pre-existing parents who influenced the birth and/or infancy of the universe. Using Hermes' principles as a framework for testing hypotheses, due consideration should be given to the idea of a conscious rebirth—as opposed to a random bang—for the creation story of the universe. That will lead to different assumptions, not only about a reality outside the universe, but also about what we are doing in it.

For example, if the universe is governed by the same principles as all levels of life, once its growth phase ceases, it should enter the aging part of the cycle, to end up back at a point to be recycled. The Principle of Correspondence would predict an entire universe experiences cycles like those of its components: birth, growth, decline, decay, and rebirth. In such a context, the event now considered a Big Bang would be more accurately described as a Grand Rebirth, involving a yin/yang merging of powers that might appropriately be considered the "Grand Couple." (See Table III.) For the ancient Hindus, the Year of Brahma is 311 trillion years long—a period which represents the expansion and contraction of the cosmos. Its conclusion is not seen as the end to time, but rather as the end to only one cycle of cosmic breathing. Consistent with the Principle of Rhythm, the Hindu traditional wisdom may yet be proved valid by Western science.

Table III

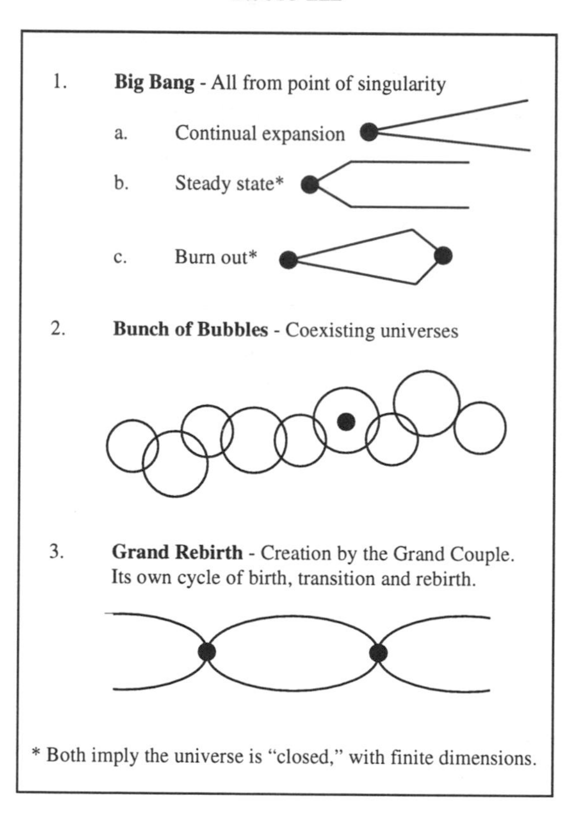

Ideas that have been cast aside as too metaphysical must be re-examined as more logical explanations of observed reality than those offered by the limited assertions of conventional science.

Although theories about origins and endings of the universe may not seem relevant to our current lives, at least one theoretical question may directly affect our experience: Are there parallel universes—the so-called

Bubble or Bang Controversy? Most scientists accept, as do most religions, the idea of a moment/point of singularity for our universe, but the debate is wide open regarding the possibility of multiple or "bubble" universes. (Such multiverse theories could account for some of the gaps in current scientific explanations.) If they exist, some scientists believe conscious beings may be able to experience them through "worm holes" that serve as a gateway between parts of one universe or between separate universes. If multiple universes exist and humans can experience them, what kind of beings are we?

Some scientists argue that, even if other universes exist, we cannot "know" them because of the boundaries (termed an "event horizon") of our own. In what may be only a question of semantics, others hypothesize that our universe is really a branch, located somewhere as only part of a larger system. (Bubble theory advocates point to characteristics of the universe such as Hawking radiation and homogeneity to support their view.) If consciousness is cosmic, then conscious beings may not be bound to one universe. This flexibility could account for many of the so-called paranormal and psychic phenomena discussed in later chapters. Since we can only "know" with our ordinary senses what has happened in the five-sense universe since its beginning, we cannot "know" that which occurred before or outside it. But perhaps there are other ways of knowing. If information and consciousness can travel faster than the speed of light (not bound by mass), then perhaps we can learn what has happened outside this universe.

Whether the multiverse theory can stand depends, according to some, on the existence of a GUT that explains everything for our universe. Right now the apparent indestructibility of the proton is one factor holding up the proof of a GUT. If the decay of this particle can be proved, some scientists believe they can be assured that all four forces collapse into one. On the other hand, the proton's indestructibility might prove that the universe is inherently unstable, and therefore subject to quantum manipulation by cosmic forces. While the search for a simple theory of everything continues, we can draw new insights by taking a different perspective (as detailed in Chapter 4) and reenvisioning the universe we now imagine.

Astronomers are beginning to collect evidence that stars are not only dying but are being born. Stars are not necessarily always products of gas clouds collapsing into superbright concentrations of energy that are visible to us through their infrared wavelengths. The Hubble Space Telescope (HST) has now revealed that stars are apparently being created from the clusters of energy formed from colliding galaxies, energy that is equal to 500 billion suns. The HST has revealed that the blue star Eta Carrae, previously thought to be fading into oblivion, is in fact erupting.[8] It appears that stars, like other organisms, are born and then die, but before they die they join violently with other stars to produce offspring that perpetuates the stellar family. The same may be true, if the Principle of Correspondence is operative, of entire universes on a larger scale. If universes die and are born again, is a process of conscious reincarnation at work?

Consciousness Enters

With the mention of consciousness, the description of our universe takes a dramatic turn. We know consciousness exists because we have it. Consciousness is more than thinking: in consciousness we are aware of our thinking. Yet few physicists attempt to confront the everyday reality of human self-awareness because it cannot be perceived directly by the five ordinary senses. They subsequently ignore evidence of so-called paranormal abilities, including the anomalous results of telepathy and psychokinesis that show specific characteristics of a fifth and/or sixth force.

As human beings, we elaborate our individual experience of consciousness through the physical senses, but we are keenly aware that we are more than they reveal. In this book, as we review the ways in which individual and group consciousness affects matter and energy through forces focused by human intent, it will become evident that a larger consciousness is at work around us. Even though most scientists personally recognize the inconceivability of a universe such as ours occurring by chance, the profession's norms discourage formal inquiry into external consciousness as causation. (See the suggested reading list at the end of Chapter 9.)

Some scientists, who admit the impact of consciousness but have no theory about it, conclude with a concept Brandon Carter has called the "anthropic principle." Carter's theory, drawn from quantum physics, holds that we as human beings create the universe by the way we look at it, just as an experimenter who wants to measure light finds a wave of light where another could observe a particle. Unfortunately, the circular logic of this concept permits one to escape without fully addressing the problems of primary conscious intent: Where did conscious intent come from? How does it work? Does it have to inhabit matter at all? Another theory, that of hyperspace, which posits up to six dimensions beyond our four-dimensional version of the universe, leaves room for a "scientific incorporation" of consciousness into a descriptive model.

Any comprehensive theory of the universe must take into account the role of mind or consciousness. Hawking and others believe our everyday world can be accounted for by a limited number of basic physical laws, like a theoretical spider web amid which nature is delicately hung. Thus far their formulation of laws is not as comprehensive as the Hermetic Principles, which do provide for conscious intent in the patterns of creation.

This chapter has thus far focused on the objective universe, from the outside looking in, with a selective interweaving of the Hermetic Principles that combines physics and metaphysics. From now on the text presents a more integrated viewpoint, but before we leave the macrocosmic scale we revisualize the entire schema, seeing the discernible elements interacting as one integral whole.

The ground of being is beyond our knowing: it is the nothingness (from our perspective) that is outside space-time. From the ground of being* that existed before the Grand Rebirth of our universe, came conscious intentions, which shaped cosmic energy into waves and particles that, with the arrow of time, produced the four dimensions[9] of which we are a part.

From those particles and waves came not only galaxies and star systems, but self-directing creatures. As in the forming of galactic families, these creatures now compose their own groups and make up untold numbers of species. Among these beings are the Solarians—including humans—of our sun system. We do not know the origin of this creative consciousness,* but we realize we are part of it. Our prime questions are: *why* and

* Interestingly, "ground state" in physics means that energy is as low as it can be, and "ground of being" in theology (Tillich) refers to the realm from which creation derives.

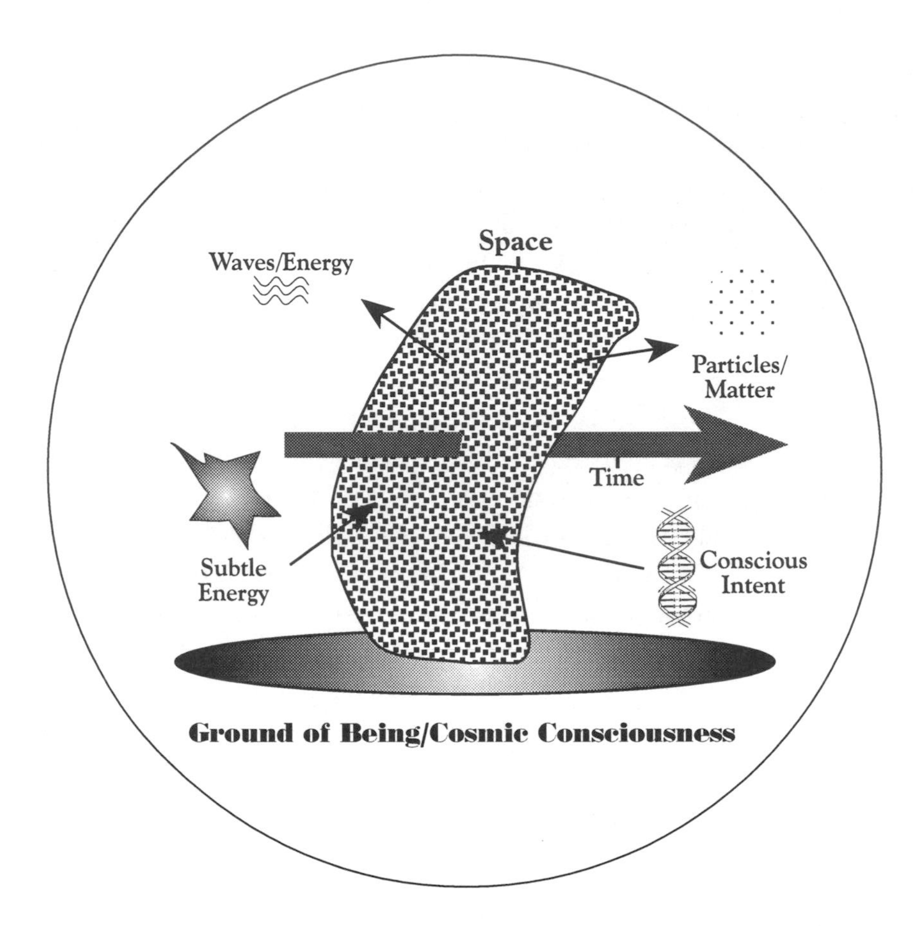

how did such self-aware beings come to be? I dare not tackle the question of why, but I believe it is possible to understand something of how. By focusing initially on the macrocosm, we can use the Principles of Gender and Polarity to infer something of our birth. We can use the Principle of Correspondence to better understand the dynamics of our development.

To borrow from traditional Chinese cosmology, potential is the yin of this fertile consciousness; expressive intent is its yang. This play between potential and intent can be observed at all levels in the universe, from galaxies

---

* Some people use words like "soul" and "intelligence" synonymously with "consciousness." I prefer "consciousness." "Soul" has too many mystical, value-laden connotations; "intelligence" has become too tied to the concept of IQ.

to personal development. Our galaxy has at its center two huge forces—celestial foci that are analogous to the cosmic Grand Couple. They emit several million times as much radiation as our sun, and are a strong source of radio waves. They may be comprised of a dense, fiery group of stars and/or a black hole. These two bodies on a galactic level correspond to the fields created by the acts of the Grand Couple on a cosmic scale and of two human lovers merging their polar energies in mating. At each level, the result is creation of offspring (an appropriate word given the energetics involved).

Celestial bodies analogous to the human couple are at all the interstices where the material plane meets with fields of subtle energy and conscious intent, which is required to focus the other two realms. These intentions are expressed through energetic patterns known as morphogenetic fields.[10] or the energized precursors of objective structures. Whenever consciousness merges foci of intent to create with subtle energy, individual entities result in the material realm—whether stellar, human, or atomic. There is no reason to believe that cosmic consciousness is less involved in the creation of children than in the creation of star systems. The inherent cosmic principles that precede and animate our universe determine the framework for, and the character of, this creative intercourse. In other words, conscious artists of creation at all levels are working with the same available media and within the same cosmic principles.

From such an intuitive but logical model, it is reasonable to hypothesize that conscious intent impregnated the subtle energy womb of the Grand Couple, from which came the physical universe with its polarities of energy and matter—or *matenergy*, as I prefer to label the spectrum. Out of this fusion of intent with *matenergy,* through the medium of cosmic energy, comes the world of spirits, atoms, molecules, and organisms. Our universe is still vibrating from the kinetic energy generated by pressure in its passage through the Grand Rebirth canal. It is the varied combinations of different vibrations of *matenergy* that give rise to the four classical elements—earth, fire, sky, water—*and* plants and animals. They all derive from the gender and polarity of *matenergy* and consciousness. These dynamics of creation are delved into more fully in Chapter 4.

Though human beings are discovering more about the universe each day, all that we have learned constitutes little beyond the knowledge that we are part of it. We know that we are subject to being consumed at any moment by forces of cosmic proportions. We are beginning to understand that the same forces drawing scores of stars larger than our sun to the brink of transmutation are leading humans on an equally adventuresome journey. We appear to be readying ourselves for dramatic phase shifts. Though we may seem to be flying blindly, like birds in the falling snow, we have an inherent awareness of direction. The following chapters help us get a better grasp of that human capacity.

While still unsure of the origins of our species and how we came to reside on this planet, we feel an impulse to return to the stars. Our current generation has achieved moonfall, demonstrating to ourselves the capability to sail off the planet. We have sent unmanned missions to photograph other planets and are preparing, in erratic steps, to land on Mars. Our probes and signals now reach beyond the solar system, and by the next decade we are likely to have an orbiting space station that will serve as a base for planetary exploration. Involving Americans, Russians, Europeans, Asians, and others, this era of cosmic exploration will dwarf any physical achievement by modern man. However, frontier science and a new metaphysics shall unveil the role of consciousness in the microcosm and our larger history, achieving breakthroughs that will outstrip even the advances in interstellar exploration. Expansion of our inner and outer horizons will result in a rediscovery of our true nature, and our Solarian Legacy.

NOTES

1. Chatwin, Bruce. *The Songlines* (Penguin Books: New York, 1988).
2. Darling, David. *Deep Time* (Delta of Dell Publishing: New York, 1989). Most current estimates of the age of the universe range from 15 to 20 billion years.
3. Three Initiates. *The Kybalion: A Study of the Hermetic Philosophy of Ancient Egypt and Greece* (The Yogi Publication Society: Chicago, 1912).
4. Capra, Fritjof. *The Tao of Physics* (Shambala: Berkeley, California, 1975).
5. Thorne, Kip. *Black Holes and Time Warps: Einstein's Outrageous Legacy* (W. W. Norton: New York, 1994).
6. Hawking, Stephen. *Black Holes and Baby Universes* (Bantam Books: New York, 1993).
7. Halliwell, Jonathan J. "Quantum Cosmology and the Creation of the Universe." *Scientific American* Dec. 1991: 76-85.

8. "Stellar Vision." *The Sciences* Mar.-April 1994.
9. Ouspensky, P.D. *Tertium Organum* (Vintage Books: New York, 1982). The arrow of time used here is neither the subjectivist (perceived by us) nor the absolutist (eternally apart from us) view of time, but a statement of entropy or other organic progressions inherent in the phenomenal realm.
10. Sheldrake, Rupert. *The Presence of the Past: Morphic Resonance and the Habits of Nature* (Collins: London, 1988).

Chapter 2

Microcosm: Creative Void

From where do the complex, but elegant particles come that make up the beautiful life forms, including humans, who inhabit the universe? Physical science takes us on a microscopic journey from cells through molecules and atoms to a shimmering world of subatomic particles. Each refinement of our instruments reveals smaller and smaller bits of matter, jostled about by unseen, but measurable forces. Why are they there in one nanosecond, and gone the next? Where is home when they're not here? What creative power brings them into being? What is the explanation for why humans seem to share that power? How does consciousness, which seems nonmaterial, master energy and matter?

◆ ◆ ◆

HAVING COMPLETED A BROAD SWEEP through the universe's 20-billion-year history, it is now time to explore the other end of the spectrum—the microcosm of tiny subatomic particles that make up matter. To consider their nature, shifting out of nothing and back to nothing again, requires a mental leap into the void, where patterns are created by consciousness that have the power to shape apparent nothingness into tiny quanta of something. The terms *void* and *nothing* are not meant to be synonyms for *space* and *vacuum*. Space is not empty: it is filled with particles and waves. When we remove matter from space to create a vacuum, we leave the energy of wave fluctuations. Therefore, by void and nothing I mean states of nonform and nonmateriality, beyond space-time: the realm of concepts or ideas—the *logos* of the Greeks.

Space-time is composed of three dimensions or directions, plus time. In space-time any event can be located, if only approximately, at a specific point for a given instant. Even though some scientists are speculating that there may be as many as ten or more such defining dimensions, they are all variations of some local observer's space-time. The void, on the other hand, works with principles and through the medium of forces that are not known to be subject to the rules of the physical universe.

In seeking a metaphor to help illumine the nature of the void and its relationship to ordinary reality, I have chosen the familiar "looking glass" from Lewis Carroll's *Alice's Adventures in Wonderland.* The looking glass is a useful device (also used by John Briggs and David Peat) because it is understood to represent a world that is the obverse of this one, a world potentially less distorted, where analogous, but different principles operate. It represents a simultaneous existence that can be temporarily visited from this one. The nature of a looking glass allows for the fleeting appearance of objects with a translucent quality, an attribute of virtual reality. Additionally, the metaphor captures the essence of the microcosm, where the initiative can come from either side: events beyond the glass can result in entities and actions on this side.

Dealing with quantum mechanics and space-time in the microcosm requires us to grasp the concept of creative activity devoid of energy and matter as we know it. At this point we have come full circle in the Einsteinian concept of relativity: the infinite nature of the macrocosmic universe is not different from the infinitesimal nature of the subatomic one. They both arise from something we can only describe as nothing. As we saw in Chapter 1, the concept of a Big Bang, or something from nothing, is illogical. Therefore the context from which the universe and its microcosmic parts arose can only be considered nothing for purposes of our five-sense reasoning. We leave the discussion of the cosmic womb from which the universe was born to a forum beyond this book.

Most scientific theorists in the twentieth century—an era organized by the counting of time and the measuring of space and matter—have fully vested themselves in the material realm of the ordinary senses. Theorizing the void goes beyond even ephemeral images like a photon of light, a structureless particle that is so near to nothing that it easily blinks in and out of existence. So some scientists now postulate a force that shapes amorphous vibrations and waves into neutrino quanta that manifest no matter, yet may comprise ninety percent of the mass in the universe. Though such concepts of this material reality are elusive, they are pedestrian compared with the flashes of insight that are necessary to envision the creative workings of the void.

Building Blocks

Before leaping into the void, it is useful to review how scientists have searched for the building blocks of life. Matter has successively been revealed to be broken down into smaller, more elusive bits. For a long time people thought of distinct elements as the irreducible constituents of matter. Then philosophers hypothesized and scientists proved they were made of atoms. Students learned an atom was the smallest indestructible piece of reality (a trillion million of them fit on the head of a pin). But atoms in turn were revealed to be made up of electrons, protons, and neutrons. Theorists thought the building blocks had been figured out, and they categorized elements by the ratios of protons to neutrons and grouped them according to the number of electrons in the outermost shell of the atom. Then several combinations of even smaller particles were discovered within each atom. But the breakdown did not stop there.

Physicists now have taken us into the world of even smaller bits of what many believe is the frontier of the physical realm.[1] To converse about phenomena at the bottom edge of the barely visible world, the concept of quantum, which means "how much" in Latin, is used. In the physical universe everything appears to happen in quanta—the progressions, or leaps, from one level or state to another with nothing in between. This smallest unit of measurement signifies matter or energy cannot be broken down into another unit: additional force would transform it into nothing. The following review shows how far the search at this level has taken scientists.

The smallest quanta that have been broken down and still perceived through mechanically enhanced senses fall into four families: Leptons (with six particles), Quarks (with six particles), Clasons that include photons and gravitons, and Weakons that include “W” and “Z” particles. Since all particles seem to exist in complementary pairs (Principle of Polarity), the number in any family is always equal. Therefore, as only five quarks have been found, it is assumed that for the “down” quark, a sixth or “up” quark must exist. The latter is a most tenuous bit of matter, believed to exist for a mere fraction of a millionth of a second. Researchers theorize it exists, but have not found one in almost twenty years of searching and trying to create one artificially. In early 1994, the Fermi National Accelerator Lab near Chicago, using an accelerator firing protons and antiprotons at nearly the speed of light, reported *traces* of what they inferred was the “up” quark. The same laboratory later reported that smaller particles, subquarks, may exist. Other hypothetical subatomic particles, although as yet unproved, have been given names like neutralinos, squarks, selectrons, and axions.

Given scientists’ dependence on mechanical devices to perceive such tiny particles, some believe some subatomic particles may be only artifacts of the technology. Therefore, it is difficult to determine if we are measuring something as it exists or if it appears in that form because of the way we measure it. This is an important issue when the factor of conscious intent is introduced in the conversion of subtle energies to ordinary phenomena. If that conversion process is operative at the atomic level, it makes the subatomic level irrelevant for purposes of conscious creation. A grasp of the process of materialization of matter and energy from the void requires identification of the level or levels at which it occurs. Therefore, further review of some research within the current subatomic taxonomy may be helpful in that identification.

Quarks, making up the neutrons and protons that form the nucleus of an atom, always act in triads. (The triangular shape provides strength to structures at many levels of life.) Leptons counterbalance the quarks, making up the subatomic particles that revolve around the nucleus. These two categories of matter (Quarks and Leptons) are counterbalanced by two families of energy (Clasons and Weakons), with two members each.

Analysis of these four members (gravitons, gluons, photons, and “W” and “Z”) provide the underpinnings for the four basic forces of the universe discussed in Chapter 1. Of the four, conventional thinking considers that all except the graviton have their effect inside the atom. The graviton, a hypothesized vector boson (a ball-like concentration of energy), believed to exert the pull of gravity, has not yet been detected with current devices. An alternative view, holding that gravity may be an electromagnetic-like charge, places it alongside the other three. The gluon (aptly named) is a vector boson that holds the nucleus together. The photon boson carries the electromagnetic force, and the "W" and "Z" bits of energy effect radioactive decay. These four represent the reciprocal forces of cohesion and disintegration that characterize the two tendencies of all matter and energy. Momentary manifestation in space-time occurs when the two tendencies are in balance.

Some scientists postulate other bits of matter—"virtual" particles or fields—but as of yet we have no instrument to measure them. At this very microcosmic level, the dividing line between a particle and its unmanifested pattern is very fuzzy, hence the term "virtual." Nevertheless, this conceptual plank begins to bridge between our material existence and the void: it introduces the possibility of a medium of interaction. For example, Belovari of Hungary[2] believes discord in a body's "virtual photon field" could wreak havoc on the physical organism (analogous to the idea that damage to the auric field results in ill health).

The particles of quarks and the bosons of energy, and their anti-matter counterparts, come into being out of this "virtual" state beyond the arena of mass and energy as we commonly know it. This state could be the "primordial stew" of Chapter 1, or an aspect of the intermediate state I label *energeia* (discussed in more detail in Chapter 4), but neither is the void. While the process of their awakening is obscure, once these virtual quanta manifest in the phenomenal realm they can be converted and reconverted. Light can be a wave or particle, energy can turn to mass and vice versa. Quantum physics offers insight into such points of conversion from one form of energy and matter to another within the physical realm. But we have to look to metaphysics, or a *metascience*, for insight into the question of how materialization derives from the virtual state. How, for example, does emotion leads to neuropeptides and invisible fields change water to medicine.

The answer seems to involve counterparts of matter and energy in some conductive medium between physical entities and the void. The label *energeia* recalls such ancient concepts as "chi," "prana," and "mana," or "zero point energy" in modern physics.[3] In this arena, as in others, different terms are used to describe the same phenomenon, but they all point to such an energeial link between the subatomic world and a context called the void.

Looking-Glass Void

Changing from the microscope, peering at the matter-energy spectrum, to a metascientific telescope, diffuses sight into a larger context, the void from which the whole universe arises. Looking out there we end up with the same "nothing" we found in the microcosm. It becomes clear that the personality of the universe is akin to the personality of an individual: it has no matter. At this metaphysical level, all that is there is *logos*: ideas, tendencies, and potential patterns. *Logos* is a field no more substantial than cosmic memories, yet it is more powerful than galactic-scale lightning converting energy to particles, or black holes swallowing star systems. *Logos* pre-existed our twenty-billion-year-old universe and will still exist when the universe collapses back into the undifferentiated cosmic egg, gestating for yet another cycle.

If we are to comprehend what is now clearly the next frontier of conscious awareness, we must engage in an intellectual dance between form and nonform, between that which *is* (*matenergy*) and that which *is not*. Unfortunately, our language is not conducive to this type of discussion: by saying the void from which something arises, implies that the new something comes from something before. Perhaps the Muslim belief in an "uncreated Koran" and the Buddhist tenet, "Form is emptiness and emptiness is form," can help us illumine the difference. In what might be called a Zen paradox, we have to use our mind to clear our mind of all previous concepts in order to make this imaginary leap into the void. It is possible we will discover William Blake was correct: looking through the glass at our invisible soul, we find that our material self is the perfect reflection of it.

The Buddhist concept of a fully seeded *shunyata* helps one to understand that from the void comes everything: it holds infinite possibilities for accidents or chance as well as conscious growth. The void provides for either deliberate or de facto development, with the looking glass frame establishing the limits of freedom of choice for an incarnated being in the space-time of this universe. Constrained by, but not prisoners of the mirror and its frame, humans are conscious co-designers of their own being. The very existence of the capacity for self-awareness empowers them to cross over and participate in the ongoing design of patterns in the obverse world of the looking glass. Within this framework, human powers enable individuals to change much of what is, and, therefore, what is to be, but they cannot retroactively undo what is already done. In understanding their limits humans gain their freedom.

One must keep in mind what the void is not: it is not dark matter (which does not emit or reflect electromagnetic radiation) or antimatter (hypothetical particles whose traces have been observed for 40-billionths of a second in the European Laboratory for Particle Physics and the Fermi National Accelerator Laboratory). Some scientists link dark matter and antimatter together as the unknown factor in the total-mass equation of the universe, but they are two different things. The Principle of Polarity would predict, and many scientists assume, an equivalence (in terms of substance and force combined) of matter (light and dark) and antimatter in the physical universe. So the void involves a field more elusive than either antiparticles or virtual particles.

In his book *The Holographic Universe*, Michael Talbot[4] suggested that—at a level beyond the current concepts of physics—matter, energy, and consciousness blend into a single field. But combining that triad still leaves a range of unexplained influences and communications experienced by human beings. To explain the ability of the human mind to interact with and influence ordinary matter and energy at a distance, it is necessary to invoke another medium—the realm of subtle energies. It may relate to or act on the above-mentioned "virtual" fields. More comprehensive treatment of this concept comes in Chapter 4. For the moment it suffices to illustrate the indirect impact of consciousness on matter at the microcosmic level.

Mind as Catalyst

Researchers are accumulating evidence of the power of an proactive local mind to shape matter. For example, Deepak Chopra's elegant and powerful description has popularized research that demonstrates the human ability to think neuropeptides into existence.[5] (Neuropeptides serve as messengers that carry instructions to activate cell functions, including those of the immune system.) Celeste White has summarized significant evidence to indicate that individual human mental effort can have a minute, though vital, impact on the creation and/or manipulation of crucial bits of matter. Her selection includes: methods to reduce the electrical excitability that triggers epileptic seizures; guided imagery to affect the immune system, blood flow, and heart rate; and use of hypnosis to cure genetic illness (warts).

Robert Becker, in *The Body Electric*, shows how the expression of feelings or thoughts causes direct current (DC) to flow along the body's nerve sheaths (perineurals). In this instance, the mind shapes waves of energy, parallel to its previously described influences on particles of mass. These examples of physiological and biochemical reactions show that the two basic categories (matter and energy) of physical building blocks are susceptible to the power of one being's mind.

Applying the Principle of Correspondence, one can assume that scaling up the mental/emotional effort (by many people thinking together)* likewise scales up the physical effect. The ratio curves are yet to be determined; however, researchers should be able to extrapolate to entire organisms or systems from the measurable influence that a small effort has at the cellular level.

The above research offers important proof of the creative force of consciousness working with unknown subtle energies to shape the world of physical phenomena. Chopra calls the interstice between active intelligence and silent intelligence the "gap." In the looking glass metaphor, the gap is the glass itself, the point before potential becomes manifest. However, silence exists only in the perceptions of the physical senses. In the realm

* This is the principle through which received prayer, or other conscious focusing of intent, affects the course of individual or group life.

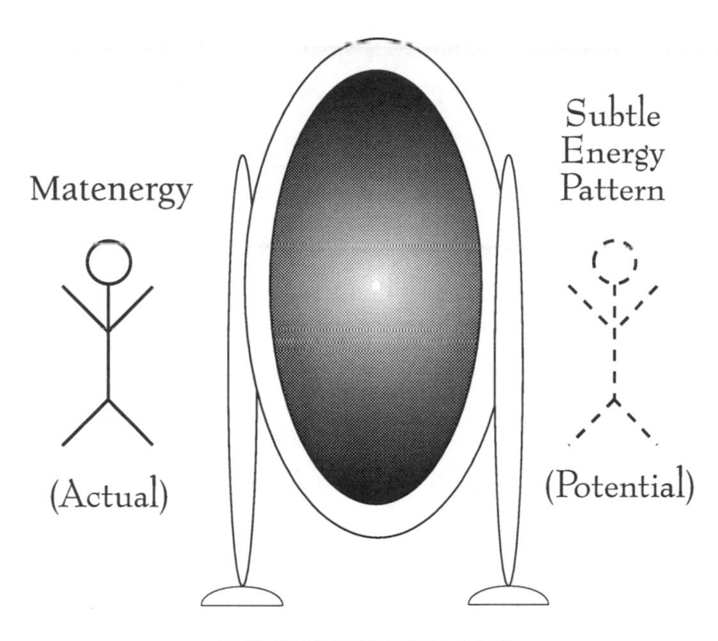

LOOKING GLASS

beyond the glass, there is apparently a lively reality of subtle energy, vibrating harmoniously and in total complementarity to this side.

One difficulty with the looking glass, or any other physical analogy, is that it still perpetuates the idea of a dichotomy, like that of spirit/matter, brain/mind, or heaven/earth. We must somehow be able to see ourselves, the mirror and frame, *and* the reflection, as a whole within an even larger void.* We need metaphors, like the Zen moon-in-the-water concept: the water is the subject and the moon is the object, and "moon-in-the-water" is a field jointly created by the apparently separate subject and object. But it, too, is part of a larger field encompassed by our mind.

Even the process of selecting a metaphor reveals the requirement for some overarching element, like conscious intent. Such an element is

---

* A friend rapidly deteriorating from HIV-induced decimation of her immune system was fading in and out of consciousness. One day I gave her a black and white photograph I had taken and carefully developed. Seeing it, she sighed and asked how I knew to offer it to her. "That's exactly how I have been feeling," she remarked. The next day she departed this incarnation. The photo was of a basket of kittens looking at themselves in a mirror: it was impossible to distinguish between the furry, playful animals and their reflections.

necessary to account for the choice of one from many available options. It is conscious intent that has the power to collapse infinite possibilities to a point of decision or manifestation.

Conscious Intent

The "river of creative consciousness" continually shaping matter and energy never ceases. We paddle our canoe in it, taking advantage of eddies and whirlpools, but we cannot stop the stream whose origin lies in the unfathomable ocean. So the question for conscious beings is how to express intent in the continuous flow. The answer lies in the small choices people make, where there is freedom to choose. To pose no resistance to the current is as much a choice as direct action. The Principle of Cause and Effect is equally valid in moments of silence or nondecision. With a choice one can manifest a different emotion (a form of subtle energy): happy or sad, the choice creates a neuropeptide out of the available stock of the hydrogen, carbon, oxygen, and other atoms in our brain cells. Thus a thought, while not redirecting the river, initiates a reaction that affects the body's course. The challenge is to identify where the paddle of individual intent can be inserted in the flow of life.

French scientist Jacques Benveniste, in research on homeopathic medicine, proved that a liquid continues to carry the residual pattern of an antibody dissolved in it, even after the liquid has been diluted to a point where no physical trace of the antibody can be found.[7] His research indicates that pattern traces appear to have an existence independent of their material manifestation. These pattern traces differ from the holographic pattern, which resides in fragments of matter. While we do not yet understand the mechanism, such pattern traces can be projected by computer-generated light waves or by human thought. Rupert Sheldrake's description of morphogenetic fields may explain how thoughts can have such power. These fields, described as "thought bundles" by Nancy Parker in her fine novel *Omega Transmissions*, contain the constructs or patterns necessary for manifestation in the physical realm.

Through both trace patterns and morphogenetic fields, matter is influenced by ideas—whether latent memories or newly created ones. A latent memory trace may exist from a creative moment outside our space-time, or be the result from a long-ago conscious thought developed within our space-time. (For an individual, they may be residuals from a previous life.) The newly created field can come from almost any contemporaneous source.

Given the power of the ongoing river of consciousness, if parents-to-be do not make a choice (perform a deliberate mental act at conception) to shape DNA patterning, does the memory of an earlier creative act—latent in the parents' gene structure—determine the zygote's inheritance? Under these conditions, it would seem to follow that if a genetic structure for the body-to-be was not predetermined during the original creation, and no choice was made at conception, *some intermediate creative force subsequent to the Grand Rebirth has acted.* Without having to know who or what the intermediate creative force was, one may assume every currently effective genetic pattern is the memory trace of a bygone creative thought.

Without understanding how the patterns are formed in the void, we can still experiment with their catalytic nature. If the patterns cannot be found in matter, e.g., the totally diluted homeopathic solution, there must be a mechanism for their transfer from the ideational realm to perceptible form. Whether it is in the creation of one atom or the repatterning of genes, clearly some type of verifiable energy or force is involved between the intent and the affected particles and electromagnetic waves.

Pattern as Intentions

It is important to keep in mind that "space" as we know it is *not* the void out of which matter and energy directly materialize. Space may include some sort of etheric medium[8] in which potential energy waves and matter (two poles of the feminine Yin) remain inert without the infusion of intent (masculine Yang), but the void is the realm of potential patterns. Only after one of the potential patterns is selected (by consciousness acting in the void), energized into a subtle form (in the etheric or intermediate realm), and manifested (balance of cohesion and disintegration in the realm of matter and energy) do we have a space-time event.

At this point it might be helpful to clarify possible confusion about the term *energy.* In the context of space-time, Thomas Bearden helps us to understand that we use the word "energy" to describe both the capacity for activity and activity itself.[9] He and others see the capacity as the source. But as both the capacity and its active state are in space-time, as two poles of one spectrum, many agree there is another form of energy—sometimes called *subtle energy*—the unseen and unfelt force activated by conscious intent that serves as the transforming medium. The concept of a subtle energy linking the two other aspects (physical and imaginal) of our universe implies consciousness is the highest level "organizer." Prior to either an intentional or accidental joining of matter and energy comes the truly creative event, when the idea or pattern is first conceived. This purposeful thought or conscious intent that designs the initial pattern comes from the realm of either a local mind or general consciousness.

The experience of a three-level universe that is conterminous to the point of singularity, and wholly interdependent, is difficult to model

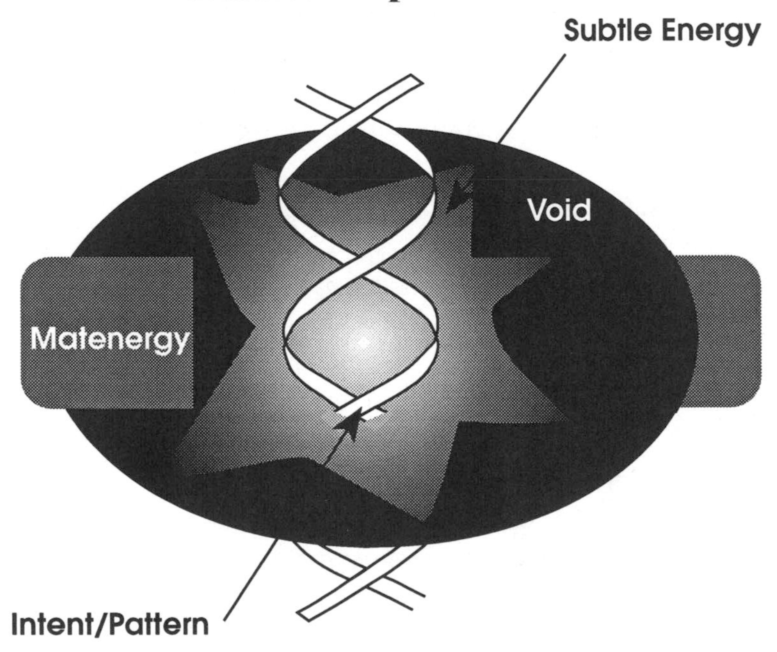

graphically. Talbot and Itzhak Bentov believe the holographic model moves closer to explaining the totality of such a multidimensional universe.[10] In a hologram one can literally know the whole from the part, a modern demonstration of the Principle of Correspondence. The smallest fragment of a holographic record can be broken off and yet still reflect the entire image when subjected to laser light. The hologram itself is a record of interference of patterns, i.e., the way an object breaks up the waves in a field of light. To make the image visible, the film is illuminated by a laser beam comparable to that used in producing the hologram. The interference patterns then redistribute the laser waves (all the same frequency), revealing the photographed object suspended in its field of space.

The holographic principle is reflected in nearly every human cell, where all the DNA instructions for the entire complex of organs lie quiescent, yet capable of giving birth to whatever is ordered—a brain, a kidney, a toe. At the same time each cell knows its own place, its own purpose. The inescapable conclusion is that the patterns for all parts exist at every level of the universe. Though invisible and unmeasurable, they make order out of chaos,[11] converting potentiality into actuality.

In this omnipresent intelligence, decisions to manifest one part or the other are communicated instantaneously throughout the whole, unlimited by matter, energy, or the speed of light. Physicists demonstrate scientific proof of this capability by separating paired subatomic particles and registering that they react in tandem when either one is individually acted upon. Similarly, biologists prove the same point when they separate human white cells at great distances from the body and track the cells' reactions to changes in the donor body's state. We see the result of the communications; we do not yet understand how it occurs.

Current theories positing explanations of this communications phenomenon include Irish physicist John Bell's 1964 assertion that the reality of the universe is nonlocal and British physicist David Bohm's belief that an invisible field connects all matter and events in the universe.[12] This field is assumed to exist outside the domain of ordinary reality, clearly part of the void in the cosmic looking glass metaphor. It acts—like the Christian Holy Spirit and the Hindu Brahman—as the breath of life for cosmic beings.

The holographic construct, to some extent, illuminates a crux of universal reality with which we must grapple (and for which quantum mechanics offers no easy explanation): What happens when something singular fragments in every direction? Recall the opening chapter's descriptions of that "incomprehensible first second." Everything that is now was condensed into an invisible point to begin with. And though we have a universe that spans billions of light years, its entirety is still wholly represented within each infinitesimal part. This again illustrates the full circle: microcosm reflects macrocosm and vice versa. But at what point does organic, self-reproducing life (in contrast to simple innate consciousness) enter the picture? How does the generic breath of life become focused in a particular entity's pattern?

Matter to Life

We can perhaps gain insight into the dynamic of conscious creation by examining the building blocks (larger than the subatomic ones discussed earlier) of organic systems. Atoms make up organic submolecules such as sugars and aminos. Combinations of these submolecules result in molecules which in turn combine to create the nucleic acids (DNA and RNA), proteins, carbohydrates, and lipids that make up living cells. The cells then link together to form tissues. Tissues make up organs, which in turn make up fully functioning systems such as the human body. All such local acts of creation must take existing elements and build with and upon them: nothing totally new is manifested in what is essentially an act of reorganization. Does this power to co-design and reorganize reside on the conscious-being continuum at some point between the Grand Couple and humans?

Louis Pasteur, who discovered how to use heat to kill microbes in food, "proved" to the French Academy in 1864 that life on Earth could not have arisen spontaneously. Earlier experimenters had mistakenly supported the idea that life could arise on its own because organisms had begun to grow in solutions from which they had not excluded airborne microbes. Pasteur proved that the organisms in those test tubes came from airborne microbes. His experiment dealt a mortal blow to Aristotle's idea that a particular mixture of inorganic particles could by chance spring into life. We are left with the obvious: seeds of new life anywhere in the universe come from somewhere else, with unfathomed direct linkages to the original cosmic birth.

Still unconvinced, twentieth-century science continues to seek the creating link between inert chemicals and life forms. The italicized conventional story in Chapter 1 recounting the development of elements and celestial bodies in the universe does not include the development of life forms. Biologists have assumed the chance formulation of life, and focused on trying to understand the process of mutation and evolution among single-cell and higher forms of life. Though generally accepted that the basic elements of life on Earth also exist elsewhere in the universe, no one has yet shown how to bridge the gap from inert matter to a single-cell organism.

Cyril Ponnamperuma at the University of Maryland and many other chemists are trying to fill the breach. One theory they examine is that extreme heat, as that found in volcanoes at the bottom of the ocean, may serve as the catalyst for the spontaneous production of bacteria. An earlier hypothesis held that lightning bolts struck the rich ocean soup and catalyzed chemical bonding, resulting in a cell that could move and regenerate itself. The applications of huge amounts of energy in various forms (heat and electrical) in many experiments have failed to prove such a hypothesis.

The fact is science does not yet know how the first cell of any particular life form is initiated. However, once initiated, cells reproduce their own kind according to patterns of DNA and RNA. The DNA form nucleotides or letters of the genetic code. This code or message is contained in the way the molecules are bound together and aligned in a long, ladder-like chain. The RNA molecules play a role in transferring and interpreting genetic messages.[13] The central point here is that the DNA, assisted by "messenger" and "transfer" RNA, dictates all chemical activity in the body—chemical manufacture, transfer, modification, and usage. The DNA is the blueprint that shapes life forms. But we are left with the unanswered question, who was the architect that designed the DNA?

Were all the DNA blueprints present at the universe's creation, to await activation according to a built-in schedule? Interestingly, researchers have identified in the genes uses for only about five per cent of the DNA. Is this other so-called "junk" DNA the basis for future development? If so, how does it become activated? Social scientist Jean Houston teaches that our

greater potential is encoded, waiting to be released by our conscious efforts. She draws on the Greek concept of entelechy (promoted by German embryologist Hans Driesch in the early part of this century) that implies an innate, nonphysical drive to self-fulfillment. This view means that all beings, in effect, are programmed to differentiate and become more complex.

Whatever the source, all life on Earth—from the simplest virus to the human being—has a common base. The human genome has approximately 3 billion individual links or base pairs* that provide the blueprint for the multibillion-celled human body. The common earthworm, by contrast, has only 959 cells (including 302 nerve cells). Although the two bodies exist on vastly different scales, the same genetic principles apply.

Is the Earthly genome the base for all life in the universe? The presence of comparable forms in meteorites reportedly from Mars indicates there may be a common base, at least in our solar system. Examination of alien beings may reveal that Earthlings have the key to the structure of all life in the universe. Reports of alien corpses, including a filmed autopsy of an alleged alien reportedly recovered from a UFO crash in 1947, may lead to some answers in this area.

Have humans finally eaten of the mythical tree of knowledge by advancing into the field of genetic engineering? Tomatoes, celery, carrots, peppers, potatoes, corn, coffee, and melons have all been "improved" to satisfy customer taste and corporate profits as a result of this knowledge. With recombinant DNA, genetic engineers can take desirable genes from one species (either plant or animal) and splice them to others. Two projects use chicken protein to make potatoes more resistant to microbial infections and Arctic fish protein to prevent frozen vegetables from turning soggy. After the cloning of Dolly the sheep, the ethical as well as the practical issues of genetic manipulation have become a public priority.

Other initiatives are underway to use genetic manipulation to cure human diseases. Retrovirus-based technology is now used to cure some illnesses by inserting new DNA between the tail and head of the diseased cell. Some believe the same can be done for HIV, the virus that causes AIDS.

---

* The Human Genome Project, funded by the National Institutes of Health, is scheduled to map the coded sequence of all these fragments on each gene by the year 2005.

Could retroviruses (they invade and modify the cell's genetic code) be vehicles for significant redesigning of the human gene pool? The 1997 announcements of successful mammalian cloning with adult DNA demonstrates the long-run limits of genetic manipulation may be constrained only by the human imagination and self-imposed ethical curbs.

While scientists can manipulate the genetic content, they do not yet understand the genetic off/on switch that results in cell differentiation, or how different tissues derive from what started as a single cell. While humans have identified the keys to reorganization of life forms, they know neither how to make the keys nor who made the original set.

Mind as Creator

The use of conscious intent is a more effective route to human participation in a creative life process, including maintaining health, than purely mechanical manipulation. Examples given earlier in this chapter demonstrate how the mind calls into existence simple matter (neuropeptides) and influences energy (the flow of current). Others explain how genetic engineers technically reshape matter at the subatomic and molecular levels. It remains to explore the possibilities for human use of general consciousness for more direct creation. Modern research evidence has been found to support the conclusions of many cultural traditions that human consciousness communicates with and influences the behavior of other life forms.

Long before Cleve Backster hooked his interspecies communication-testing polygraph equipment to the leaves of a *Dracaena massangeana* (a common house plant),[14] the Kahunas of Hawaii teach people to ask the desired food for its permission before harvesting it to eat. They believe plants and animals have feelings about their purpose and function in the food chain, and appreciate being asked and honored before being eaten. Generations of experience have taught that plants taken home after a sincere request live and flourish, while those unceremoniously ripped out do not. Believing an omniscient consciousness extends to rocks as well, they do not haul them from their place, knowing the spirits of the rocks can make

trouble.* Similarly, the Native American tradition of asking permission of the game before hunting it and of vegetation before its gathering, then thanking them before eating, is rooted in an appreciation of both the physical connections and the conscious channels that link all living species. They too believe the foods we consume have feelings that deserve expressions of respect.

Backster's research has confirmed the energetic response of living cells to their imminent incorporation into another being is muted when humans tell them in advance what is to happen. He has found that spontaneous acts by the researcher to boil an egg or eat yogurt results in intracellular agitation, measurable by electronic sensors. Conversely, he has found that the researcher's expressing the intention in advance has a "calming" effect on the food cells. Such examples of the impact of thoughts or intentions on matter illustrate the flow of some level of communications between the human mind and organic materials.

Assuming there is an ongoing, reciprocal flow of such behavior-influencing communications among local concentrations of mind (animals, plants, and individual cells), how does the conscious being deliberately intervene in the natural flow to bring about a desired end? The answer—at this point an intuitive one—involves a clarity of focus grounded in definite emotions. The process appears to work in a manner analogous to the progression from "gas" through "liquids" to "solids," i.e., from "amorphous" through "evolving" to "definite."

A vague new idea falls into the "amorphous" category. Once this idea "evolves" into a "clear" concept, its coherence can give focus in the realm of subtle energy. There, following the energeial three-step progression, the "potential" image enters a state of "becoming." When it is "definite," the idea is then "actualized" in the material realm. One can observe the three progressive small phases in each of the big steps: idea (to conceptualize), motion (to energize), and form (to actualize).** These three steps will be central to the overall model of reality developed in Chapter 4.

---

* The local post offices reportedly receive many returned rocks with plaintive notes asking that they be taken back to the mountain from which they came.

** Ayurvedic philosophy divides all existence into three analogous doshas (categories): Vata (thought), Pitta (initiative), and Kapha (endurance).

Table IV

Three Steps in Creation		
Conceptualize	Energize	Actualize
Idea	Motion	Form
Amorphous Evolving Clear	Potential Becoming Definite	Space Energy Matter

An example illustrates the progression: After someone had the first "clear" idea of a candle, it entered the "potential" phase as soon as there was real intent to create. The movement to gather materials placed the pattern in the "becoming" phase. The "actualization" of the idea was ignited by fire. The same sequential process applies to all fields of human life: agriculture, food preparation, health, psychokinesis, sports, politics, or economics. The idea of democracy first starts with clarity about a few basic assumptions. Only when a number of individuals attach emotional support to the ideas does democracy have real potential. That emotional energy translated to action results in the practice of democracy. A vision of health must be underpinned by emotional commitment in order for the cells to get the message to do their part and for the individual to eat appropriately. To facilitate the bending of metal through conscious intent, one focuses the idea on the metal, the metal's atoms become agitated, and when as a result the material softens, only a slight pressure will bend it.

Recognizing the power of this natural process, it is foolish to assume specific limits to creative powers exist in natural law until they have been tested and re-tested. Currently perceived constraints may actually be due to false or limited interpretations of cosmic law. If a clearly focused thought is more than a fleeting mind game, emotionally energized to a level of potential, will it be realized? Or, are there certain thoughts that cannot be actualized in space-time? When we learn the answers to these questions, we will realize our fuller power as cosmic beings.

The characteristic of space-time existence appears to be that mind and matter are inseparable, bonded together by a force field that literally shapes the material into a form reflective of the mental pattern. The illustration below, from modern art and phenomenological psychology, depicts the inseparability of the three aspects of reality: at the microcosmic level, mind in this space-time is always manifested as some perceptible form or energy.

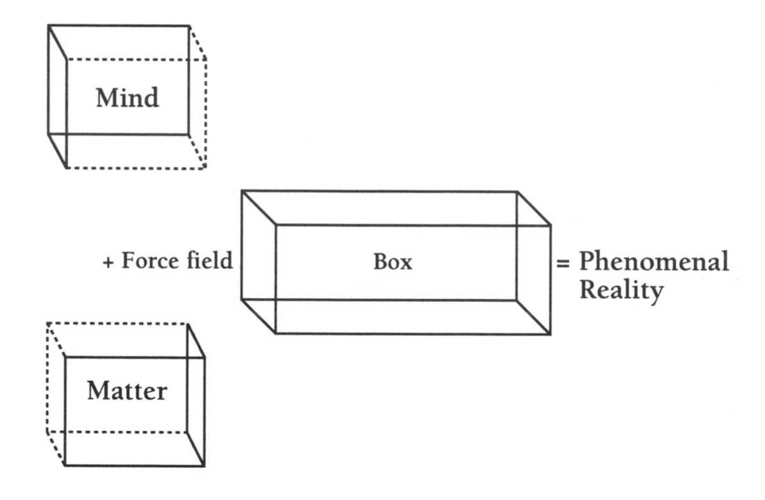

While the nature of each manifestation may be of infinite variety, providing for much interesting research, we must first unravel the intrinsic process if we are to consciously manage it.

Nature's Limits

A previously mentioned experiment at the subatomic level may demonstrate an as yet undefined medium—variously referred to as *chi*, *ether* or some other subtle force*—that affects matter. When physicists split a

* In 1775, Franz Anton Mesmer labeled the unknown force "animal magnetism" or universal life force, hence, our current use of the term *mesmerized* to indicate the use of an unexplained force.

pair of photons and project one to the opposite end of an accelerator, their manipulation of one photon is accompanied by an instantaneous counteraction in the other, i.e., the negatively charged particle is manipulated and the positively charged particle reacts in a complementary fashion. It is as if both exist in some form of hyperspace where distance is immaterial, a phenomenon currently unexplained by conventional theories. It points to a subtle force field parallel with matter/energy—but not bound by ordinary space-time—that could account for the instantaneous transfer of information in many so-called paranormal experiences.

The research of Cleve Backster (reported by Robert B. Stone)[15] shows the same magic (i.e., still unexplained) at work between matter separated from human beings and their subsequent thoughts. Backster has conducted experiments in which white cells taken from a person and transported miles away react to changes in the mental state of the donor. By simultaneous monitoring of the activities and emotions of the donor and the electromagnetic activity of the stored cells, Backster has been able to chart significant correlations of the latter with the former. The energy or communication flow is instantaneous, with no apparent attenuation of speed over distance. Such thoughts may travel faster than the speed of light, which for decades has been assumed to be absolute. That hypothesis could, using Backster's research protocol, be adequately tested on interplanetary trips like the Mars probes.

Given the research highlighted here, and similar work, it is now justifiable for a prudent person to accept that mind does communicate with and influence matter at the subatomic and cellular levels. Through the mechanism of thought, consciousness or mind likely shapes reality in more ways than we can currently conceive. It is important to remember, however, that human consciousness—individually and collectively—can shape microcosmic reality only within limits, due to certain characteristics of our phenomenal realm. One of these limits appears to be the direction of the stream or arrow of time.

Some assume that if there is a "prime mover" there must also be a "prime motion" characteristic to the universe. In our space-time universe the direction in which microscopic particles move (flow of time) determines an important aspect of their nature. For example, a particle can be either a positron or an electron, depending on which direction it is traveling. A positron becomes an electron when it is forced to move in the opposite direction. This seems to indicate that we cannot arbitrarily change the nature of a phenomenon while a particular directionality is maintained. To change a vector (velocity combined with magnitude and direction) of anything requires the application of a greater force. For example, light travels in a straight line unless it is bent by a great star—or other powerful center of gravity—that warps the light of local space-time. If humans were able to apply enough force, perhaps they could affect the flow of time, but not without counterbalancing consequences (Principles of Polarity and Correspondence).

Only a power beyond our universe can mitigate the influence of its inherent arrow of time. Physicists have speculated that worm holes or superstrings may make it possible for some matter to travel faster than other, thereby having the effect of going backwards, but these theories do not dispose of the argumentation set forth here.

Another constraint on the power of mind appears to be the innate mortality of ordinary matter, even at the most elementary level. Each particle, atom, and element has its own cycle, from Alpha to Omega, after which a re-birth is necessary for a new cycle to commence. These life cycles range from a nanosecond for subatomic particles to eons for star systems. Each entity in the universe, as a function of its specific circumstances, has a built-in time to expire: each particle, entity, and being has its own rhythmic pattern from birth to death (Principle of Rhythm). For example, the human life span appears to be related to the duration of the Earth's rotation around the sun. The implication of biorhythms and other patterns is that the power of any level of consciousness less than that which created the universe must operate within inherent constraints. Discovery of the constraints operative in this stellar neighborhood would make it possible to infer the degree of power its beings have for conscious co-creation.

If the mere expression of an individual's intent to be joyful can create molecules, what could thousands of united minds accomplish in one powerful expression of emotional energy sent coursing through animals, plants, and ecosystems alike? As the mental/energetic effort is scaled up, the physical effect is likely to move up proportionately. By judicious extrapolation it may be possible to calculate just how much prayer is necessary to heal a malignant tumor, or how many meditators are required to bring peace to squabbling factions.

Another constraint is the reverse flow of influence of matter on consciousness. The ingestion of certain substances into the body's cells can influence the person's state of consciousness, as can the removal of substances on which the body has come to depend. For example, a deficiency in serotonin has been related to an increase in depression or anxiety, and the ingestion of melatonin appears to induce sleep. The conscious being has to continually deal with such material influences on its mind and energy (recall the box figure above).

All activity in the universe, then, appears to be subject to certain inherent constraints, yet some of those constraints may be more malleable than we have thought. Some frontier physicists are questioning the immutability of Newton's Law of Gravity, speculating its obvious effect may be a function of charge rather than mass.[16] People should test and re-test each perceived barrier to their use of mind: we are likely to learn that many barriers fall by the wayside as consciousness expands.

Cells to Systems

The polar opposite of "to constrain" is "to support or strengthen." The space framed by these two forces provides for creation or development. Energized by those polarities, the universe has a built-in impulse to increasing complexity, offset by one toward disintegration. The combination of constructive and destructive principles informs the growth and differentiation of organic systems from cells. Assuming the Principle of Correspondence, better understanding of the dynamics of physical systems will likely lead to more effective approaches to behavioral issues.

The most elementary form of organic life appears to be at the cellular level.* A living cell is a chemical factory of life, becoming so through the ordering power of DNA. The pattern of the now-famous double helix[17] determines what *kind* of cell the molecules will organize themselves into. The

---

* Science currently dates the first micro-organism on Earth to around 3.85 billion years ago.

cell can be a self-sustaining organism, or it can be a part of a larger system fulfilling a specialized role.

Following its internal guidance, a cell, comprised of nucleus, cytoplasm and other organelles, is able to take its place as part of a larger whole—tissue in plants and animals. The tissue in turn is but a subsystem of the organs for which it serves as building material. These organs join to make possible living organisms, including humans—the most complex organism of which we are now directly aware. The facilitating principles appear to be conducive to more and more complex beings, suggesting it is likely there are beings more complex than humans elsewhere in the universe. Since the material building blocks can be rearranged in infinite patterns, the degree of complexity is likely to be more a function of the underlying consciousness than physical constraints.

Spectroscopic and other measures of light, radio waves, and particulate radiation show the Earth's elements and compounds are not unlike those elsewhere in the cosmos. Physics and chemistry that appear to be the same here and there lead to the tentative conclusion that the same creative principles may apply to all parts of the whole.

The living cell is a microcosmic manifestation of all the Hermetic Principles. Each has a definite rhythm to its life cycle. Its maintenance of boundaries, the internal dynamics of its components, the intake of new matter, and its reproductive mechanism have a striking correspondence to the lives of other living systems, including star systems. The responsiveness of cells in their service as building blocks for organs and larger living systems demonstrates the working of cause and effect. Cell behavior illustrates the omnipresence of vibration. Its subparts clearly show the working of polarities, and gender is operative throughout cell regeneration.

Thus, we can learn much about the whole as we focus on any of the parts: the microcosm leads us directly to the macrocosm. For example, study of an anomalous interaction in particle physics (separated photons reacting simultaneously) leads to insights into the execution of psychokinesis (thoughts affecting matter at a distance). Both involve instantaneous communication through unknown channels. Discoveries in either area could also help increase knowledge of the working of successful prayers.

Ironically, by focusing on the microcosm of the atom and the power that can be unleashed at that level (nuclear weapons and heat for generators), humans have been led to even more profound awe before the heavens. But even in such an unfathomably powerful universe, knowledge seekers find themselves capable of entering into a mood of playfulness. New knowledge can only encourage people to engage consciously in the cosmic dance between thought and form, being and nonbeing, to experiment by using one's self as a portal into unknown realms.

If these reflections on the interconnections between levels of life forms and the influence of transcendent principles throughout creation is correct, one can extrapolate from those dynamics known to operate and use them in testing new hypotheses. This puts humanity in its own particular place of power. The unique impact of human consciousness on the phenomenal realm is made feasible by our "in between" character: We are more than simply evolutionary extrapolations from a lower animal order, yet neither are we total masters of our own fate. Better understanding of the microcosm will result in a more secure sense of our limitations and freedoms.

Microcosm as Model

From the perspective of the Hermetic Principles, we can apply insights from the microcosm to many aspects of life. An illustrative example of the correspondence between levels is the behavior of atoms roughly simulating the actions of heavenly bodies. In an experiment with the "Rydberg atom" in which electrons are artificially placed far away from the nucleus, the orbiting electrons behave like planets. Such examples do not definitively prove the Principle of Correspondence, but they do provide the basis for such a hypothesis. Attention to the quantum, molecular, and cellular levels has provided sufficient evidence to make connections among the realms of *matenergy*, subtle force, and consciousness. Having split the atom in the era of fission, humans can now focus on extrapolating micro-insights to larger individual and societal issues, taking the path of integration of knowledge and experience to an era of fusion.

The modern scientific focus on the microcosmic level—the trees instead of the forest—has resulted in great advances in physics, chemistry, and biology. Now with a shift in focus to connecting microcosmic fundamentals, we can begin to perceive patterns that we did not know existed—Gregory Bateson's "patterns that connect." This period of masculine (yang) expressionism has produced knowledge which has fertilized and prepared the global mind for the next phase of feminine (yin) gestation. An interim period will see the rebirth of a local yin/yang balance, a new flowering of understanding and insight among humans about their current home planet. Decades, if not a century or more, of the yin cycle will be required for humans to fully assimilate these new discoveries. They can enter then another period of yang expression that could propel them among the stars of this galaxy and perhaps beyond. (The Principles of Gender, Rhythm, and Polarity are involved.)

Each intellectual discipline takes its turn in bringing new expressions of insight. While the arts emphasize certain media of expression and the sciences focus on others, all disciplines follow an inherent pattern of rhythm, going from one polarity to another. A flourishing in science and the arts in the fifteenth and sixteenth centuries appropriately was termed a "renaissance." That rebirth involved the rediscovery of knowledge from the polarity of an earlier period, which was then synthesized with fresh insights. The intellectual (birthing) pangs in many fields, brought about by their inability to provide adequate answers to current problems, indicate industrial society is about to enter an era of a "new renaissance." Perhaps it will be one of *metascience*—an approach that flies unfettered by the dogmas of the past, and propels the minds and souls of earthbound beings into the cosmos.

There is an unfailing impulse toward wholeness within each person. A friend of mine, who at age fourteen realized her oneness with all of creation, has spent the rest of her life trying to explain it and elaborate it through literature. William Blake had his experience of unity at age four and sought to communicate it to the rest of us through poetry and engraving. The search for a unifying yet comprehensive concept of reality has characterized the efforts of both mystics and scientists for millennia. Renee Weber[18] captures the essence of that commonalty among several leaders who dealt with the issue in either the material or metaphysical realms. Plato thought perfect forms were the inner fabric of reality. Newton believed his Law of Gravitation unified all masses in the universe. Maxwell identified the unity of magnetism and electricity. Albert Einstein experienced his transcendent awareness through an intuitive encounter with the reality of relativity—a truth that has eluded his scientific peers. He elaborated on its implications for several decades and spent the last thirty years of his life seeking to unify *matenergy* and space-time into a singular source.

An entire generation of physicists has followed in Einstein's footsteps, seeking to integrate their abstract concepts into the Grand Unified Theory. Although the search may be premature, before all mutually exclusive but complementary principles have been sorted out, their impulse to do so is a natural human trait. We perpetually seek our individual and collective answers to the puzzle of Humpty-Dumpty: How do we put the pieces together again?

A recent example of such prematurity is Steven Weinberg's *Dreams of a Final Theory.*[19] A Nobel prize winner with two other physicists for demonstrating how two forces of nature (weak force and electromagnetism) were unified in an early phase of the universe, Weinberg believes that the current scientific paradigm will soon reveal all the fundamental laws of nature, included in some beautiful yet fixed "theory of everything." Unfortunately, as is clear from the mind-over-matter, telepathic, and other so-called paranormal experiences of ordinary life, the materialistic and deterministic theories of Weinberg and his colleagues omit a large part of our reality. Any effort to formulate a unifying theory must account for the role of consciousness at the microcosmic level, described in this chapter, and the as yet unidentified media through which it works its way in matter and energy.

Practical Implications

The admonition of Alexander Pope—"The proper study of mankind is man"—now assumes even greater relevance. It is within ourselves and in our conscious interaction with the universe that we employ all the elements necessary for exploring the connections among consciousness, the subtle energies, and *matenergy*, and their relationship to space-time. We are the most obvious subjects at the present time for testing hypotheses explaining the effects of consciousness on matter and vice versa. Our challenge is to study ourselves in an integrated way, beyond the current divisions of scientific and academic disciplines. The purely physical scientist leaves out consideration of consciousness and therefore the vital base of the triangle of comprehensive knowledge. The mystic or spiritualist eschews a powerful medium of communication with nonbelievers when he or she refuses to subject personal hypotheses to scientific scrutiny.

Although physicists today attempt to dissect the moment of the Big Bang, they hesitate to theorize about the nature of that which existed prior to the explosion of the single point. Modern metaphysicians are more willing to speculate, hypothesizing that matter depends on consciousness; that consciousness, acting through some force, gives rise to subtle matter; and that subtle matter (perhaps in an energy state) converts to denser matter. Ancient wisdom, such as the Hermetic Principles and other esoteric concepts, assumes the reconciliation of these different facets of experience lies in the perception of a singular form of consciousness. This book suggests the unity may best be conceptualized via a comprehensive merging of science and metaphysics into a *metascience*.

The marriage of the outer testing of physics (a mystical endeavor itself) and the inner questing of metaphysics through human self-exploration and experimentation could lead to unparalleled insights into the nature of this universe. Gaining an understanding of the human capacity to think thoughts that create matter will begin to reveal the dynamics of the Old Testament, *logos-based* theory of creation of the material universe.

Experiments using small doses of individual and group consciousness in hypothesis-testing interplay with natural events and with other beings will identify the extent to which humanity is truly a co-creative force in the evolution of the universe. For example, disparate experiments at Princeton (Robert Jahn and Brenda Dunne)[20] and the Fortean Research Center in Lincoln, Nebraska[21] demonstrate the impact of intent on the behavior of material objects. The Princeton work provides mainstream scientific evidence of the ability of humans to have an impact on the behavior of metal balls in a computer-controlled experiment in randomness. By the expression of intent, individuals and small groups can cause statistically significant deviations from the normal distributions of balls in the apparatus. In the Fortean experiments, the ability of subjects to manipulate small objects at a distance (psychokinesis) is measured by the standards of traditional science. Widespread application of these findings to other fields will contribute greatly to expanding the frontiers of knowledge.

When basic principles are discovered at the microcosmic level, practical insights easily follow. The field of health is a good illustration. Instead of misdirecting billions of dollars in public resources into externally based chemical and mechanistic medical research, we would be wiser to study the internal cellular changes wrought by the human mind in the dynamics of the self-healing process.

More than 20 years after the passage of the National Cancer Act—an official declaration of a "war on cancer"—we still do not know what causes cancer or how to stop it. New types develop and spread. One out of three Americans now alive will develop a malignancy. Melanoma, a skin cancer, is becoming the fastest growing cancer in the world. (Perhaps if we had begun by calling it something other than war, we would have conjured up less resistance, less of an enemy.)

In search of explanations, some hypothesize that ultraviolet rays, unfiltered due to ozone depletion, shrivel DNA.[22] Others see the roots of growing cancer rates in oxygen deprivation or electromagnetic field anomalies. Yet, in a healthy human body, the immune system routinely kills cancer cells as they develop. Problems arise only when the immune system goes awry. Since we do not know how the immune system works, we do not know why it fails. The matter-to-matter approach of allopathic medicine includes antibody research (attaching killer cells to cancer cells) and toxin delivery systems (matching toxins to the cancer cell receptors.) Yet metastasis—the spread of cancer to other parts of the body—more often than not outwits these localized cancer "fighting" efforts.

Chemo-prevention, finding anticancer chemicals in food that naturally feeds the cells, is receiving increased attention. It is showing some success because natural systems spread the effects of food more efficiently than do therapies alien to body processes. While there is some progress in identifying natural substances that activate natural healing systems,[23] too little attention is given to exploring the microcosmic-level interactions among the three fields whose reciprocal play determine the health of organic life: mind, body, and subtle energy. The seminal work now being done on the power of an individual's positive attitudes and conscious intent to heal oneself deserves the society's top priority, as does the healing effect of the conscious intent of others expressed through prayer, touch, and other forms of explicit support. Such practices involve the whole-body communication systems and the mind-body-energy nexus at the atomic and cellular levels described in this chapter, and avoid the attack-oriented approach of conventional treatments.

Deliberate patterns of thought energize the cells in the body to seek a wholeness that effectively eliminates the virulent symptoms. The mental message communicated to the cells is that they can and should reject this or that as an outsider. Such natural responses avoid the side effects of typical allopathic blockbuster doses which frequently do as much ancillary harm as good. Conscious intent, by precipitating the response of the whole system, enables the body to reject the disease patterns.

Thus, it is *metascience* that humans need most—research on their own survival as a species. It offers the best route to unraveling the larger questions of the dynamics of the cosmos. "As it is above, so it is below," pronounced the Hermetic. The Stoic concurred and asserted "the Macrocosm is expressed simultaneously in the Microcosm." The Hermetic would only add, "and vice versa."

NOTES

1. Morris, Richard. *The Edge of Science* (Simon & Schuster: New York, 1990).
2. Belovari, Gabor. *New Science News* Spring 1993.
3. Devereux, Paul. "Meeting with the Alien." *UFO Journal* Oct. 1995; and Puthoff, Harold. Everything for Nothing." *New Scientist* 28 July 1990.
4. Talbot, Michael. *The Holographic Universe* (Harper Collins: New York, 1991).
5. Chopra, Deepak. *Quantum Healing: Exploring the Frontiers of Mind/Body Medicine* (Bantam Books: New York, 1989).
6. White, Celeste. Consciousness and Gene Regulation." *Proceedings*. (Conference of International Association for New Science: Denver, Colorado: Oct. 1996).
7. Benveniste, Jacques, et al. Human Basophil Degranulation Triggered by Very Dilute Antiserum Against IgE." *Nature* 30 June 1988: 816-18.
8. Von Ward, Paul, with Harold Puthoff and Paramahamsa Tewari. *Free Energy Video*: *Vols. II and III* (New Renaissance Communication: Ashland, Oregon, 1994). Note the discussions of advanced physics.

9. Bearden, T.E. *The Final Secret of Free Energy*. (Association of Distinguished American Scientists: Huntsville, Alabama, 1993).
10. Bentov, Itzhak. *A Cosmic Book* (Destiny Books: Rochester, New York, 1988).
11. Prigogine, Ilya, and Isabelle Stenger. *Order Out of Chaos* (Heineman: London, 1984); Gleick, James. *Chaos* (Penguin Books: New York, 1987).
12. Bohm, David. *The Undivided Universe: An Ontological Interpretation of Quantum Theory* (Routledge: London, 1992) and *Wholeness and the Implicate Order* (Routledge and Kegan Paul: London, 1980).
13. Hazen, Robert M., and James Trefil. *Science Matters* (Anchor Books: New York, 1991).
14. Tompkins, Peter, and Christopher Bird. *The Secret Life of Plants* (Harper & Row: New York, 1973).
15. Stone, Robert B. *The Secret Life of Your Cells* (Whitford Press: Westchester, Pennsylvania, 1989).
16. Haisch, Rueda, and Puthoff. "Beyond $E = Mc^2$." *The Sciences* Nov.-Dec. 1994:26-31.
17. Watson, James D. *Double Helix* (Atheneum: New York, 1985).
18. Weber, Renee. *Dialogues with Scientists and Sages: The Search for Unity* (Routledge and Kegan Paul: London, 1986).
19. Weinberg, Steven. *Dream of a Final Theory* (Pantheon: New York, 1993).
20. O'Leary, Brian. *The Second Coming of Science* (North Atlantic Books: Berkeley, California, 1992).
21. Caidin, Martin. "Telekinesis Demonstration." *Fate* Jan. 1994.
22. Murray, Linda. "The Cancer War: Stories from the Front." *Omni* Feb.-Mar. 1993: 50-56.
23. Moore, Meecie. *The Miracle of Aloe Vera: The Facts About Polymannan* (Charis Publishing: Dallas, Texas 1995).

Chapter 3

Unicosm: Planetary Home

In the expanse of the universe, there must be natural neighborhoods for different species of conscious beings. What comprises a natural home base? How does one identify an extended family of cosmic beings and its interactions with other families? To discover a cosmic family's heritage, one looks at its legends, ancient artifacts, historical documents, current beliefs, existing science and technology, and social institutions. But to get at the more permanent reality, one also studies the blanks in the official picture: the experiences of members not explained by conventional wisdom are analyzed. Such research gives a more accurate picture of humanity's history and destiny? Can we know our true legacy?

HUMAN BEINGS ARE POISED BETWEEN the microcosm and the macrocosm, somewhere in the vibrating totality of the living universe. What is our "place" in this marvel? Unique creatures in a divine garden, especially created by an omnipotent god for an ultimate purpose? A cosmic accident, the unlikely coincidence of random mechanical and chemical forces that somehow evolved into beings capable of pondering the nature of their own existence? Figments of a divine imagination entertaining itself? A more plausible answer draws on a broad range of available evidence from the past and present: We are both more and less than we think we are.

We are less than the sole progeny of a personal god-like entity totally devoted to our quotidian and mundane concerns. The current deification of human beings (Jesus, Mohammed, Buddha, Khrisna, and others) does not appropriately honor the nature of our cosmic parentage. On the other hand, we are full members of a large and diverse community of conscious beings situated throughout local space-time in the universe. We are inextricably connected with all of them and all facets of our universe. Unfortunately, the current limitations of our habitual assumptions and our intellectual blinders may restrict us to a diminished role in the cosmos. The following pages may empower us by suggesting an expanded concept of who we are.

Stellar Neighborhood

We are first Solarians and then Earthlings. Our sun's nurturing warmth bestows life on our planetary system, through the process of photosynthesis and heat-generated chemosynthesis. Photosynthesis has produced the primary deposits of coal, gas, and oil that have made possible the rise of modern petroleum-based, techno-mechanical society. Photosynthesis in plants provides for the conversion of inorganic substances into food that sustains animal and human life that could not otherwise exist. So, even as mobile physical beings, humans, as a function of their birth, are ultimately linked to the solar system.

Beyond providing life-support, the solar system seems to shape the inner design of planetary organisms. For example, the coherence of sunlight—its irreducible dimension—is about the size of the surface of a cell; and its resolution—the shortest distance a light wave can travel before colliding with a thermal photon—corresponds to the distance between base pairs along the double helix structure of DNA. Thus, the essential underpinnings of life (the cell and its blueprints) seem to be correlated with characteristics of our supporting sun, the defining element of our stellar neighborhood. Consequently, we may infer that life inhabiting other planets and moons of our sun would likely have comparable organic parameters, although not necessarily the same physical appearance. (However, recently on Earth life forms not dependent on light have been discovered.)

Our solar history may parallel that of similar stellar neighborhoods throughout the universe. Astronomers have recently discovered, in hundreds of locations, disks that appear to be about the size and mass of our solar system. These disks are composed of dust grains and gas, suggesting that the sun-like stars have the raw material of planets. Observations of activity around the star 51 Pegasus, reported at a 1995 conference in Florence, Italy, suggest that these specks of matter coalesce into asteroids that then collide and meld into planets. Scientists think it is probable that our neighborhood originally came into being the same way, implying systems conducive to life may be widespread. If that is correct, we have many galactic cousins populating other neighborhoods.

Our solar family is likely to be a recent generation compared to those from other parts of the universe. Only about 5 billion years ago, a sun and its dark opposite—like Sirius the dog star and its unlighted twin—moved at the edge of our galaxy. Forces of gravity drew matter from the galaxy in a celestial mating dance of yin and yang polarities that pirouetted across a portion of the Milky Way. Hunks of coalesced ordinary matter fed the appetites of the dark and the light. Those closer to the dark hole were gradually sucked away through the spatial expanse and into its hungry maw. Delicately balanced in between were nine to twelve bodies gently suspended in orbits around our self-lighted Sun. The process was not unlike the seeding, coalescing, and refining of human families.

How could these globes of matter maintain their orbits and not collapse upon their own mass? Are there naturally occurring rhythmic sizes and distances from the parent sun (Principle of Rhythm)? One such principle was hypothesized by Johann Titius (1766) and popularized by Johan Bode (1772). Bode's Law held that planets should fall into proportional distances from the sun in a series of 0, 3, 6, 12, 24, 48, and 96. Using his theory, Bode postulated that a planet was missing between Mars and Jupiter, where the asteroid belt is located.

The positioning of our planet in a particular orbit, with implications for sustaining life, is contingent on the reciprocal gravitational pulls of the sun, moon, and our sibling planets. As humans, our lives are also affected by this *interdependent family* of celestial bodies. Humanity's planetary home base lies in the vortex of multiple levels of gravitational, magnetic and other forces,[1] some of which manifest in physical ways such as female menses; others are less obvious. Astrologers believe the conjunctions of countervailing planetary forces are powerful locations that affect all life.

Since Pluto's discovery in 1930, conventional astronomy has considered nine bodies to be the planetary children of the Sun—Mercury, Venus, Earth, Mars, Jupiter, Saturn, Uranus, Neptune and Pluto. From the time of Bode's calculations in the eighteenth century, researchers have speculated about the existence of another body—Planet X. Recently, scientists have identified perturbations in the orbits of several planets, including Pluto, that could be indirect evidence of such a wandering body.

In 1976, Zecharia Sitchin published the first volume of a remarkable five-volume series of books—*The Earth Chronicles*—that not only documented the existence of the mystery Planet X, but also posited its role in the creation of Earth and its human inhabitants. Sitchin's *The Earth Chronicles* and *Genesis Revisited* are based on an in-depth study of the thousands of clay tablet fragments left by the Sumerians of Mesopotamia. Predecessors to the Assyrians and Babylonians, the Sumerians had a highly developed civilization that influenced all Middle Eastern cultures, and in turn much of the rest of the world. Nineteenth-century European archaeologists "rediscovered" them when they uncovered these "olden texts" on clay tablets.[2] Even if Sitchin's interpretations are only partially correct, they call into question conventional contemporary assumptions about the origins of human beings and their place in the universe.

These tablets bore the Sumerian recordings of the teachings of advanced beings who claimed to have been from a twelfth planet.* The tablets contain descriptions of star movements that could have only come from an off-planet perspective, not from sky-watching shepherds. The Sumerian record makes a persuasive argument that a planet X with an orbital period of 3,600 years (and with an apogee six times farther from Earth than that of Pluto) collided with a former planet in orbit between Mars and Jupiter (where Bode predicted) billions of years ago. Roughly half of the satellite remained intact,

---

* Their home body — a planet way beyond Pluto, like the Planet X of modern astronomers — was labeled the "twelfth" planet because they considered the Sun and the Earth's moon, in addition to the nine planets, to be members of the stellar family.

becoming our Earth, a planet with a single dry continent surrounded by a monolithic sea. It settled into a new orbit, our current one. The other half became the asteroid belt—the "bracelet of heaven" composed of asteroids, ice balls, and millions of hunks of rock—that still fills that space and orbits the Sun.[3] (See the following graphic.) Our moon could have been near the current path of what is now the Earth and been captured by its gravity, or it

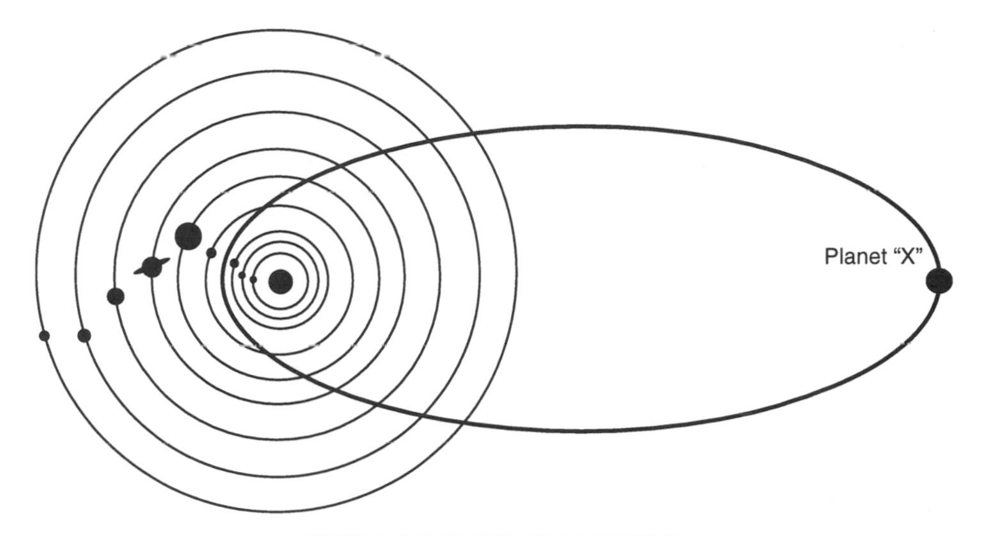

THE SOLAR SYSTEM
(Not to Scale)

could have been a piece remaining from the smashed planet.

According to Sitchin's theory, Planet X, coming from the far reaches of our solar system, not only reshaped the Earth, but brought seeds of life picked up on its long voyage.* Whether that life was in the form of simple spores (postulated by the theory of panspermia) or conscious beings, our neighborhood appears not to have been as isolated as we have believed. The Earth's interactive galactic history is more complex than we have imagined and makes for exciting hypotheses about the role of the humanoid (a human-like being) in interstellar space.

---

* Astrophysicist Brian O'Leary points out several problems with the hypothesis of such a wandering planet, given the apparent stability of all the planets' current orbits, but says the issue deserves scientific attention.

Turbulent History

The currently postulated and generally accepted geological ages of Earth are shown in Table V. But we do not yet have the whole story. The fact is that we know very little about our past. For example, legends of the Great Flood and the sinking of Atlantis—both recorded in Western historical documents—only hint at relatively recent cataclysmic events.

Table V

Earth's Geologic/Fossil Record		
Age in Millions of Years	Geological Eras	Developments
5,000 - 4,600	Earth's Birth	Oldest date of rock
4,500 - 2,500	Archean Era	Reverse RNA sequencing indicates life 3.8 billion years ago.
2,500 - 570	Proterozoic Era	Very simple life forms
570 - 245	Paleozoic Era	Movement of animal life from sea to land. First land plants. First amphibians. First reptiles.
245 - 65	Mesozoic Era	First mammals. Called Age of Dinosaurs; extinction.
65 - Present	Cenozoic Era	Current life forms. Follows extinction of dinosaurs and 2/3 of all species.

When we survey our planet's past, we confront tremendous gaps in knowledge about many cataclysmic events: the rise and fall of continents; widespread ice ages and floods; the disappearance of dominant life forms; and unexplained shattering of large areas by comets, asteroids, and radioactive explosions (like the low-altitude Tunguska River-area explosion in Siberia that leveled hundreds of square miles of forest in 1908). What are the forces that have subjected our home to such trauma? What is the likelihood of similar occurrences within our civilization's time horizon? Are conscious beings able to predict and do anything about them?* The ongoing debate between "stabilists" and "dynamists," with their opposing ideas of the degree of upheaval caused by cataclysmic events, has divided the scientific community. Have our ancestors in some way been responsible for any of these events? Is humanity currently engaged in activities which have the potential to precipitate a new Armageddon? The following brief examples illustrate the indefensibility of pat assertions about the Earth's past and justify thc posing of such qucstions.

Abrupt shifts on the planet have been documented, such as the much-discussed extinction of dinosaurs and two-thirds of all species around 65 million years ago. A near-instantaneous event is evidenced by freshly eaten buttercup flowers found in the stomach of a frozen woolly mammoth. The earth's fossil record has provided paleontologists with evidence of two other mass extinctions—one around 245 million years ago when more than 90 percent of existing species became extinct, another 11 million years ago when 30 percent were eliminated.

Many scientists now agree that a huge object (containing large amounts of the rare metal iridium) from outer space slammed into Earth about 65 million years ago, leaving a crater 110 miles wide on Mexico's Yucatán Peninsula and causing a dust cloud that could have killed off dinosaurs and other life. (The Earth has about 140 known craters of similar size; the world recently watched an analogous event as the Shoemaker-Levy-9 comet string smashed into Jupiter.) Others dismiss the impact and dust cloud theory and attribute the extinction of dinosaurs to shifts in the Earth's magnetic field or changes in the ratio of oxygen in its atmosphere. The record is by no means definitive, but theorists attributing dramatic life changes to atmospheric

---

* There is evidence to indicate that humankind not only suffered as the result of geological, atmospheric or other "natural" cataclysms, but also at the hands of more advanced beings. For example, the story of the Tower of Babel in Genesis 11:9 records that survivors from the flood — warned and instructed in the means to escape by one of multiple "gods" — were then scattered over the face of the earth.

fluctuations have some evidence on their side.

Dramatic declines in the ozone layer have impacted plant life and the weather system, which in turn have affected the magnitude of the polar ice caps. The atmosphere at the end of the last Ice Age had much less carbon dioxide than today. Pankaj Sharma, a radio chemist at the University of Rochester, has concluded that cosmic-ray bombardment was 41 percent more intense 21,000 years ago than it is at present. This suggests that the earth's magnetic field was then much weaker.

For a long time, geologists believed the Earth's mantle (40 to 2,000 miles deep) was a solid hunk of dead rock. But recent research has shown it to be composed of many types of rock and zones of varying temperature and density—a finding that indicates a more volatile planet than many previously thought. A recent innovative theory—"crustal displacement"—implies the entire crust of the Earth shifts, like peel slipping around the core of a piece of fruit. Such movements could account for the alternative freezing and thawing of ice layers on various parts of the globe.

Recall Sitchin's report that the Earth originally—when propelled into its current orbit by Planet X—consisted of a single land mass. In the 1920s, Alfred Wagener, a German scientist, postulated the theory of a single land mass. The breakup of that continent over the eons through geological, and possibly humanoid-initiated, events dramatically affected surface life. The division of a single land mass originally inhabited by one civilization might account for the universality of various art forms and common primordial, multicultural myths that now span the globe.

Ordinary forces of gravity and heat transfer, as well as changes in both mass and vibration, may have led to periodic shifting and breaking of the continental mass.* An example is the rising of the great mountain ranges of the Andes, Himalayas, Rockies, and Alps through volcanic eruptions, earthquakes, and continental drift that may have occurred near the end of the Cretaceous period, 70 million years ago. Civilization's influence on surface temperatures may be equally important in such volatility as the movements

---

* Jason Morgan's Tectonic Plate theory, put forth in 1968, suggests the earth's crust, including dry land and sea, is divided into ten to twenty plates (which are subdivided into platelets). The plates float on a layer of liquid and jostle against each other, both horizontally and vertically, at fault lines. The friction at fault lines results in earthquakes, volcanic eruptions, and the violent lifting or sinking of huge areas.

of liquids below. Over the eons the configuration of land surface and water surface has probably changed several times, including episodes during an earlier ice age (600,000 years ago) and the most recent glacial period that ended only 10,000 years ago.

Evidence indicates that relatively recent tectonic activity (10,000-12,000 years ago) submerged large parts of land connected to what is now Florida and Central America, Morocco, and the Iberian Peninsula. Perhaps the whole area now known as the Mid-Atlantic Ridge—considered to be the site of the lost continent of Atlantis—underwent just such an upheaval and subsidence around the end of the last Ice Age. There is evidence that a now lost civilization existed at the time of this cataclysm, and may have played a precipitative role in its timing. See Charles Berlitz' *Atlantis* for a cogent and plausible analysis.[4]

Core samples, dating between 10 and 50 thousand years ago, taken from the ocean floors in this area reveal animals and plants from surface life forms. Stone structures have also been photographed at great depths in this area. Similar structures were found on the Canary and Azore Islands by sixteenth-century European explorers. Other structures and roadways found leading off the shore lines of Florida, Mexico, and Belize demonstrate the sinking of land on which apparently highly developed settlements once existed.

During that same time period, with the end of the last Ice Age and the start of warm ocean currents in the North Atlantic and North Sea, three major civilizations (Egyptian, Babylonian, and Hindu) developed their historical calendars. The Mayan and Olmec calendars also have a beginning date during that period (8570 or 8500 B.C.E.). Could a singular revolution in consciousness (either externally motivated or arising indigenously) have occurred in all these regions at the same time? Were the calendars "reset" for a new cycle by survivors—now scattered over disparate areas—from a single lost continent?

There is a possibility some of these earth changes may have involved interactions with intelligent beings. Analysis of ancient evidence strongly suggests such interactions in our planetary history. Javier Cabrera has published an analysis of millennia-old pictographs found on at least 11,000 stones discovered in Peru in 1961, when an earthquake shifted the sands of the immense Ocucaje Desert.[5] According to Cabrera's interpretation, these ancient engraved Stones of Ica portray a global warming—occurring perhaps a million or more years ago—caused by a buildup of heat that could not escape an atmosphere saturated with gases, primarily CO_2. The stones depict humanoid activity contributing to this vapor shield. Resulting temperatures were so scorching that rock-like surfaces were softened and new continental configurations were fused. He also posits that some dramatic climate shifts, tilting of Earth axes, and biosystem changes can be attributed to the effects of such ancient technologies. Other evidence of the influence of conscious activity 55 million years ago in the area that is now California is published in the book *Forbidden Archaeology.*[6]

Cabrera believes the Stones of Ica document the existence of highly developed beings during a period when eight continents spanned the planet, including land masses which could account for the fabled continents of Atlantis, Mu, and Lemuria. The stones indicate that the current areas of North Africa and Europe were at one time connected and that today's Asian continent comprises three land masses that were previously widely separated. Current geological analysis of the respective rock strata support these interpretations.

We still have no idea how the forces of nature (gravitational, electromagnetic, biological, and others) combine to act on the Earth's mantle, which is literally the foundation of our continental home. Many believe we are in for another reconstitution of our habitat, that could include the melting of polar ice caps, bringing higher temperatures and higher water levels in the short term and magnetic pole shifts in the long term. The stones in Cabrera's custody may hold clues to predicting such cataclysms and their implications.

But we are getting ahead of the story. When we return to the investigations of Cabrera, we find reports of prehistoric beings who appear to have been akin to modern humans. In 1984, when researching the stones, Cabrera discovered part of the backbone (dorsolumbar) of a hominid similar to *Homo sapiens* in the sedimentary geological strata that included fauna and flora of the Mesozoic Era. Recent discoveries near the Biloxi River in Texas have revealed humanoid footprints alongside those of dinosaurs in rocks over 100 million years old. These footprints and a fossil of an index finger match those of contemporary humans. Such discoveries, combined with the planet's turbulent history, call into question theories of both Darwinism and creationism.

Fallacy of Darwinism

Most mainstream scientists still cling to an essentially Darwinist theory of evolution. They believe that natural selection accounts for all current complex life forms. Natural selection, in this view, involves random mutations (in DNA, as we now know) which are genetically transmitted to successor generations by the most procreative individuals of a species. Although the Earth's fossil record demonstrates an evolution of details contained in the forms of a given species, *no evidence exists to prove that a different species has evolved from another by natural selection.*[7] The existence of increasing diversity and complexity in fossil strata does not prove the Darwinist case.

Some evolutionists cite evidence of intraspecies mutations during Earth's long history to support their theory of the origins of radically new life forms. They theorize a long, gradated chain of development—from inert chemicals to simple life forms to highly developed ones. Yet holes in their postulated schema—and lack of evidence to fill them—call for alternative hypotheses that seem more reasonable. *Descriptive* theories that report on the visible results of unknown forces do not qualify as *explanatory* theories. In reviewing the issue, one should be honest about what has been proven and what has not.

Scientists have no definitive evidence—despite over 40 years of chemical experimentation by scores of labs around the world—that proves biological life sprang randomly from chemical molecules. Existing simple life forms do not reveal their origins, nor do they contain attributes that demonstrate they are precursors to other forms. Laboratories have artificially produced molecules of amino acids—the building blocks of proteins—but the molecules remain inert. Until self-reproducing life forms spring forth from organic molecules treated with ordinary catalysts (heat, radiation, and lightning) or from chemical formulas, the random spark hypothesis cannot be considered credible.

Nevertheless, the evolutionists start with this unproved hypothesis on how the first spark of life occurred, and then argue that the similar chemical and physical structure of cells in all earth life points to their common development from an original singular live cell. But, as we see from the continuing and parallel groups of even the most simple life forms, this hypothesis is also unproved.

Albert Engel of the Scripps Institute of Oceanography has discovered fossil evidence—the oldest to date—showing that heat-loving, sulfur-eating bacteria existed 3.5 billion years ago in South Africa. James A. Lake of UCLA has found the same type of bacteria still exists in isolated pockets such as hot water vents in the ocean floor and geysers in Yellowstone National Park. According to Darwin's theory of natural selection, all other life evolved from these (or similar) first bacteria, the weaker ones having died out as the stronger mutants took over. But Lake argues that the similarity of amino acid sequences *undermines* Darwinism, in that these three categories of one-celled organisms have remained independent of each other: prokaryotes, with no nucleus; and eukaryotes and archaebateria, both with nuclei.

Scientists disagree as to whether these bacteria types have evolved into more complex plant and animal cells. Some, including Lynn Margulis, well-known for her association with the "Gaia" concept, promote the idea that more complex organisms are really the results of symbiosis between two distinct but mutually beneficial partners. The nature of mitochondria and chloroplasts seems to suggest that photosynthetic and respiratory functions of eukaryotes merged as established functions rather than arising through evolutionary trial and error. Plants have fungi on their roots to help them get nutrients, and herbivores have microorganisms in their guts to help digest food. The nature of these interactions counters the notion of random evolution.

The Darwinists also claim as support for their theory the extended period of time available for modern complex life to have evolved. Such a focus is misleading, since the theory of sequential developments over extended time may be ill-founded. The assumption that an extremely long-term process is a requirement of evolution could have led to misinterpretation of details.

Now some scientists are postulating that cataclysmic events may have created coal, fossils, and rock strata in very short periods, thereby nullifying our current concept of geologic time. Scientists attribute an exclusive reign of reptiles and dinosaurs to the Mesozoic Era because they have dated some fossil remains for that era. *Any evidence that does not fit their assumptions about the time frame is ignored, or the conclusions reversed.* For example, the discovery of animal bones along with human bones in the Western Hemisphere has been interpreted to mean the animal only recently became extinct, not that humans lived a long time ago. Conversely, archaeologists assume objects from antiquity found in graves, rather than dating the humanoid occupants, were collected recently as antiques. Such assumptions can easily lead to underestimating the antiquity of the human species.

Analysis of trace radioactive elements in contiguous rock deposits has dated a primitive "Lucy-being"* back 3.4 million years. Those who tout a single gradated path of development have ignored the discovery of contiguous footprints—much like those of modern humans—in ancient ash deposits in East Africa and Texas (noted earlier). They dismiss the idea that a being, supposedly at least two million years younger, could have left footprints at the same time as its supposed "ancestor." But the current physical evidence, indicating the two generations lived at the same time, calls out for revision of our human story. Even within the more recent past the Darwinists cannot explain how possibly three DNA-distinct species (Homo erectus, Neanderthal, and Homo sapien) co-existed for almost 200,000 years. (See Table VI.)

The review thus far of the physical evidence has not included inferences about the evolution of consciousness, which will be discussed in Chapter 5. Consciousness cannot be proved by the physical remains available for modern laboratory analysis. Our limited means of dating fossils—potassium/argon decay rates in rocks; fission-track dating in damage to mineral crystals; and electron spin resonance counting of free electrons in solids—make it impossible for science writers to speculate about anything other than physical developments.

---

* The "Lucy" theory postulates a single female being from whom all subsequent hominids (the primate family of which *Homo sapiens* are considered the most recent) are descended.

Table VI

Selected Evolutionary Markers
(Not all-inclusive)

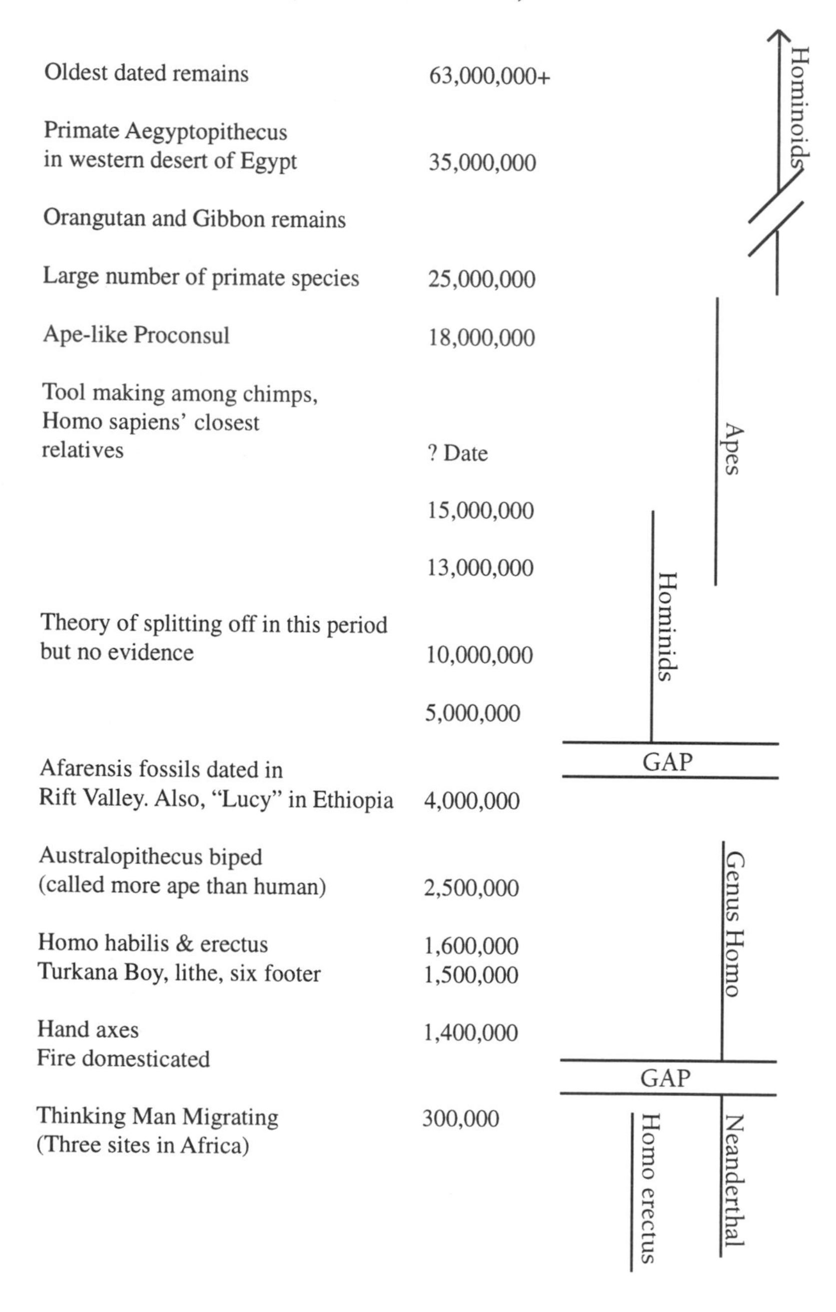

Marker	Date
Oldest dated remains	63,000,000+
Primate Aegyptopithecus in western desert of Egypt	35,000,000
Orangutan and Gibbon remains	
Large number of primate species	25,000,000
Ape-like Proconsul	18,000,000
Tool making among chimps, Homo sapiens' closest relatives	? Date
	15,000,000
	13,000,000
Theory of splitting off in this period but no evidence	10,000,000
	5,000,000
Afarensis fossils dated in Rift Valley. Also, "Lucy" in Ethiopia	4,000,000
Australopithecus biped (called more ape than human)	2,500,000
Homo habilis & erectus	1,600,000
Turkana Boy, lithe, six footer	1,500,000
Hand axes	1,400,000
Fire domesticated	
Thinking Man Migrating (Three sites in Africa)	300,000

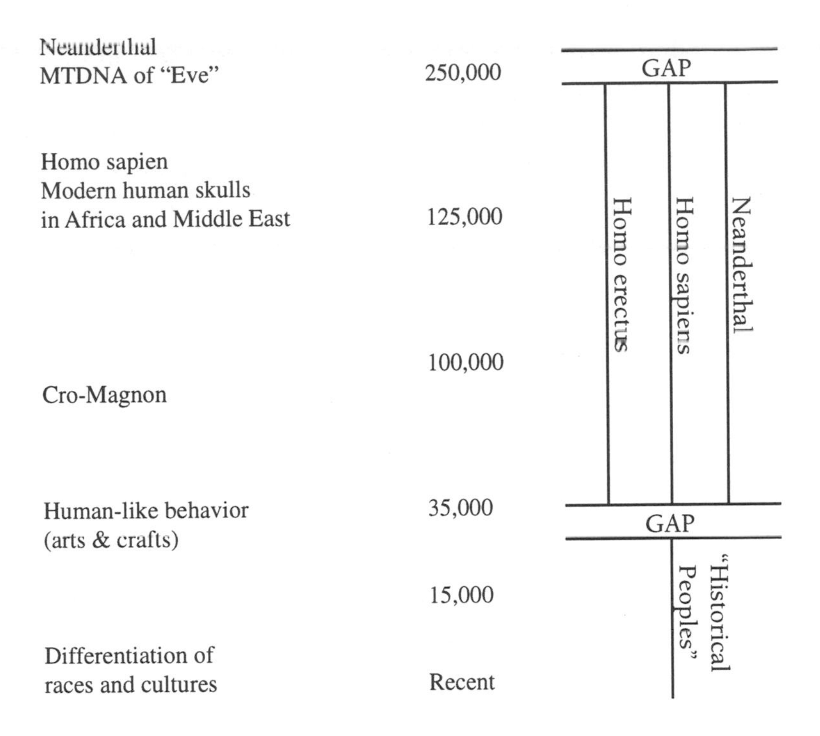

Table VI, depicting current evolutionist thinking on human development, illustrates the major unexplained gaps in the current fossil and archaeological records now being presented to the general public. It should be kept in mind that a valid theory of an inherent evolutionary process would not preclude the extraterrestrial interventions postulated later in this chapter: the two theories are not mutually exclusive.

Fallacy of Creationism

Creationism depends mainly on selected myths and legends passed down from generation to generation or inscribed in books believers hold sacred. Adherents avow that the earth and its inhabitants were miraculously given their present form by the supreme creator in the very recent past, ignoring contrary geological and fossil evidence, as well as ancient artifacts. Many of the religious texts themselves include references that support the idea that humanity has a nondivine origin, indeed that less-than-ultimate beings had a hand in human development.

Some sources describe advanced beings in various stages of experimentation using a combination of earthly materials and etheric elements.[8] According to legends of the Bushmen of Southern Africa, Cagn, the first being, gave orders which caused the universe and its beings to appear. In the cosmology of Heliopolis, men and women were formed from the tears of the Egyptian god Atum. In Old Babylonian myths, Nintu, the goddess of the Earth, molded mankind (Lullu, the savage) out of clay and animated him with the blood of a slain god. The Zuni believed all beings of Earth came from the lying together of Awitelin Tsita (Mother Earth) and Apoyan Ta'chu (Father Sky). For the Quiché Maya, mankind resulted from several phases of trial and error by a Creator and a Maker. The aborigines of Australia believe their ancestors were created by wandering beings from the stars.

Obviously these myths should not be accepted literally, but neither should they be dismissed out of hand. The remains of highly developed human settlements that predate the Biblical era deny the validity of the prevalent Christian creation theories. When one adds the kind of evidence included in this book and in geologic records compiled by mainstream science, no thoughtful person can ignore other possibilities. Both historic and current research require a re-framing of the schema for how we look at all the data.

It is plausible that advanced civilizations predating our historical era arose indigenously on Earth, then fell, with only threads of their knowledge preserved in the surviving populations. Artifacts or ruins that fall outside the recognized history of civilization are not necessarily evidence of extraterrestrial visitations. Some of these remains (the Great Pyramids in Egypt, the ruins of Machu Picchu in Peru, the Easter Island monuments, fallen cities beneath the Atlantic, etc.) could be explained by the existence of earlier human civilizations that were destroyed through cataclysmic Earthly events like those described earlier.

However, during the last few decades alone, thousands of cases[9] of interventions by nonhuman beings in the lives of people have been reported, suggesting the possibility that they are presently involved with Earthly life and intervening in its development.[10] Current abduction reports and new assessments of the origins and purposes of various historical artifacts are compatible with many of the primitive creation legends. It is quite plausible that the creationists are confusing the work of beings who are simply more advanced than we are with the notion of divinity.

An accurate history of humanoid life on planet Earth is very likely some *combination* of natural development and externally conscious intervention or technical assistance. Better understanding requires a synthesis of the creation myths, the gap-filled record of the evolutionists, the reports of genetic engineering by people abducted by aliens, and the analyses of various artifacts.

External Intervention

In the early 1970s, Erich von Däniken gained notoriety with the publication of *Chariots of the Gods* and *Return to the Stars* (later published as *Gods From Outer Space*, which sold millions of copies). Free of the constraints of institutional censorship, Däniken surveyed the evidence for the existence of advanced technology on the Earth more than one hundred thousand years ago, along with the presence of extraterrestrial beings. He speculated that these beings intervened in the evolution of hominoids (family of primates) through genetic manipulation. Since then, many other serious books and papers—some described in this book—have offered supporting evidence and analyses.[11]

Some writers—including Javier Cabrera and Zecharia Sitchin—stand out with very convincing interventionist arguments. Sitchin has amassed considerable archaeological, linguistic, cultural, and scientific evidence that supports the idea that Earth* was colonized 450,000 years ago by the Annunaki, a race of beings from a distant planet. These alien settlements are purported to have existed around three great river systems (the Nile, the

* The name of our planet adds credibility to Sitchin's intervention theory: a word linguistically similar to "earth" is used in many languages: *Erds* (German), *Erthe* (Middle English), *Ereds* (Arabic), *Erd* (Kurdish), *Eretz* (Hebrew), and *Ordu* (Persian). These all appear to derive from the Sumerian or Akkadian term E. RI. DU. The Persian word *Ordu* means "encampment" and E. RI. DU means "house in faraway built" — very appropriate if given to our planet by visitors from another home base. It is of equal interest that E. DIN was the Sumerian name for the god's home. Is this the basis for the Biblical Eden?

Indus, and the Tigris-Euphrates) and in Southern Africa. According to Sitchin's translations of the Sumarian texts, the Annunaki bioengineered the species *Homo sapiens* approximately 300,000 years ago. Thus, the first human civilizations arose in the same areas.

William Bramley[12] and Arthur Horn[13] present systematic and comprehensive reviews of currently available evidence to support the view that extraterrestrials (ETs), acting as "custodials," have manipulated human events in order to perpetuate social and political conflicts for their own ends. Their historical arguments are persuasive, even though particular interpretations of linkages among specific events cannot be proven today.

Many historical traditions report visits by beings from elsewhere in the cosmos. The oral traditions of the Dogon tribe in Mali describe visitors (amphibious space travelers) from the Sirius star system who taught the knowledge of civilization. They informed the Dogon ancestors of a dark star (maybe the black hole of modern physics) companion to Sirius. (This was publicized 50 years before its "discovery" by modern astronomers.) The Hindu Vedas describe the avatars (gods incarnate) of Vishnu as half fish/half human (perhaps arising from their space craft after splashdown). Genesis 6:2 and 6:4 speak of "gods" who consorted with the daughters of men. Does this refer to the time reported in Sumerian tablets of widespread social and sexual intercourse between the "gods" and the men and women of Earth? Modern cases, reported in large numbers, include various levels of such interactions between ETs and humans.

Most evolutionary scientists have operated on the assumption of "gradualism"—the theory that evolutionary changes occur gradually over eons, but some scientists now articulate a hypothesis that leaves room for an external intervention theory. American paleontologists such as Stephen Jay Gould and Niles Eldredge have promoted the concept of "punctuated equilibrium" where, aperiodically, dramatic shifts occur very quickly. Although they attribute such short periods of rapid change to natural phenomena, it is possible that such bursts of evolution could instead be the result of conscious, external intervention. For example, no proof exists that the three-level human brain is the result of evolution. Scientists observe its structure, then *assume* it was due to progressive mutation and reinforcement, but that is still a hypothesis. Now, genetic engineering technologies at least demonstrate the plausibility of exogenous influences.

In the last decade or so, scientists have discovered the mechanism through which deliberate intervention could have been accomplished: recombinant DNA—the result of splicing a strand of DNA into a different DNA molecule of a particular cell. This is not unlike the farmer cutting a limb from one fruit tree and splicing a similar section from another fruit tree onto the first. Scientists are now aware of the existence in humans of deactivated retroviruses that are available for DNA experimentation. It is conceivable such retroviruses were used in the redesign work of the Sumerian-era extraterrestrials. Advanced beings would be far ahead of us in this and other gene-manipulation technologies.

Mars Factor

People have long suspected our sister planet Mars of having supported earlier life forms. The numerous valleys and channels on its surface appear to have been carved by running water in a manner similar to that which occurs on Earth (Principle of Correspondence). Although Mars today appears to have no liquid water, its polar ice caps consist of frozen water and carbon dioxide. Large areas of permafrost may also exist. Whether Mars at one time supported fully conscious life has been hotly debated. Given the previously mentioned evidence of historical and current space travel, the notion of conscious inhabitants on Mars—an idea embraced by several esoteric traditions—does not seem farfetched. Rudolf Steiner's cosmology,[14] for example, holds that human predecessors lived on Mars and other planets in earlier phases of spiritual, physical, and intellectual development.

Mars is currently the source of considerable interest and controversy, due to speculation about life forms in a Martian meteorite and photographs taken by NASA's 1976 Viking orbiter. The pictures reveal formations that resemble a human face and others that resemble pyramids.[15] Scores of reputable independent scientists have extensively analyzed the Mars photographs. These analyses, reviewed by Stanley McDaniel,[16] make a persuasive case for additional investigation of the hypothesis that the photographed objects are consciously designed artifacts. Unfortunately, NASA has taken a disdainful attitude and is reluctant to treat the subject seriously.

During the planned 1993 orbits of the Mars Observer vehicle, NASA scientists could have re-photographed, with 40 times greater resolution, the Cydonia region containing these features. NASA's official commentary on this mission claimed that the Observer malfunctioned just as it was entering a Mars orbit and was therefore incapable of obtaining pictures.

An earlier incident during a Russian photo mission indicates there may be another explanation. The Russian photographic vehicle Phobos, in a 1989 mission similar to that of 1993 Mars Observer, also failed to obtain photographs. The final frames captured by the Phobos before it ceased transmission reveal images that appear to indicate it was under attack. Could it be that Mars is still an active base of operation for some beings? If so, how do they fit into our history?

Extended History

The "official history" over the past two or three centuries holds that significant human civilizations only came into being about 4,000 B.C.E. along the Nile, Tigris, and Euphrates Rivers.* In this view, not until about 8,000 B.C.E. did humans begin domesticating animals and living in villages, and not until much later did they begin extensive farming, widespread ocean exploration, and development of mechanical power. A number of pioneering researchers have pointed out serious misunderstandings resulting from this limited assumption.[17] Discoveries, like those below, could validate the existence of extended civilizations.

The Great Pyramids of Egypt have been seen as products of the period of the pharaohs, dating from about 3,000-300 B.C.E. The Great Sphinx was believed to have been created during the Fourth Dynasty, making it about 4,500 years old. Now Robert M. Schoch, a Boston University geologist and anthropologist, estimates it to be at least 8,000 years old. Using sound waves

---

* Archaeologist Anna C. Roosevelt has amassed evidence, including very sophisticated pottery, that indicates the Amazon River valley was the site of a rich civilization similar to the great river valleys of the Nile and Ganges. It is not totally inconceivable that the rich diversity of the rain forest was the result of ancient human (or, at least, conscious) management that bred useful species for food, drugs, and other social and economic needs.

to analyze the weathering of the statue's base below the current surface, Schoch theorizes that the Sphinx may have been carved from a single piece of limestone remaining in place as rock was quarried from around it. This adds credence to claims that link the pyramids and Sphinx to the ending of Atlantis, around 10,000 B.C.E.* Examples like this are slowly eroding the mindset of conventional academics who have too quickly dismissed the possibility of civilizations more than a few thousand years old.

We return now to the extraordinary work of Javier Cabrera, who discovered the juxtaposition of humanoid figures with long-extinct plants and animals on prehistoric pictographs carved in the Stones of Ica. The ancient stones show varied and sophisticated levels (some even beyond today's frontiers) of science, technology, and social development. They also contain parallels with the most ancient finds in other parts of the globe, notably Egypt and Sumeria.

Ranging from 15-20 grams to 500 kilograms (1100 pounds), the Stones of Ica have representations from the Cenozoic Era (beginning 65 million years ago), the Mesozoic Era (over 180 million years ago), and the Paleozoic Era (over 400 million years ago). Since the carvers would have had to coexist (or possess modern geological insight) with flora and fauna of those eras, it is likely that *conscious beings, in some form, lived 400 million years ago*. (The oldest known *Homo sapien* remains accepted by conventional science are now estimated to be less than 300,000 years old, but those dates keep getting pushed back with new analyses.)

Fossils only recently discovered in the geological record confirm the ancient knowledge of biology inscribed on the stones. The carvers were obviously familiar with the internal organs of the animals depicted as well as their external appearances. According to Cabrera, the stone carvings also portray maps of the cosmos, a zodiac, a calendar, planetary maps, continental maps, instruments for study of the cosmos and the microscopic world, machines for launching flights, advanced surgical techniques (organ transplants) and implements, animal and human embryology, parasitology, ritual dances, and musical instruments.

---

* A problem with dating stone is that it cannot be fixed in increments smaller than about 50,000 years, which means many ancient monuments could be much older than we have speculated.

Composed of andesite, the stones are covered by an oxide patina (indicating they are ancient) that also covered the etchings. Determining the age of the stones would require stratisgraphy and paleontology, permission for which has been denied Cabrera by Peru's Patronato Nacional de Arqueologia since July 1970. The threat such revelations would pose to that nation's Catholic theology and political institutions (not unlike the situation portrayed in James Redfield's best-selling *The Celestine Prophecy*) is apparently a deterrent to scientific investigation.

There is considerable additional evidence of such humanoid antiquity. In 1973, Hindu anthropologists reportedly found "human-like fossils" in Mesozoic rocks dating back 230 million years. The video referred to at Note 17, *Mysterious Origins of Man*, revealed the discovery of small, symmetrical stone balls containing a machine tool groove that dates back 2.8 billion years.

In his book *Atlantis*, Charles Berlitz presents evidence of technology more than a million years old:

- a silver chalice found in 1851 near Dorchester, Massachusetts, in granite rock that may have taken millions of years to form.
- a gold screw found in similar granite near Treasure City, Nevada, in 1869.
- a gold thread product and a gold animal figure with gears inside found near Cocle, Panama, in rock over a million years old.
- a godlike figurine with a metal core of wire and ceramic parts—discovered in the ancient rocks of the Coso Mountains of California.
- 716 stone-like (cobalt) disks, carbon-dated as 12,000 years old, discovered in ancient graves in the Bayanb-Kara-Ula mountains of Tibet by a Russian, Vyacheslav Zaitsov. The disks had hieroglyphic symbols and grooves spiraling from a center hole. When cleaned, they vibrated as if charged with energy.

Cabrera's book lists reports of additional astounding artifacts:

- an atomic battery found in Gabon that stopped working 100 million years ago.
- electric batteries dating from prehistoric Iraq.
- synthetic fibers in ancient Chinese burial grounds.
- magnifying glasses found in prehistoric sites in Egypt, Iraq, and Australia. (That the glasses were ground with cerium oxide (which can only be produced through electrolysis, a process requiring powerful generators) points to a technologically advanced civilization.)

Such artifacts could only have come from earlier advanced civilizations, either off-planet or home-grown. Mainstream scientists dismiss or ignore these anomalies because they cannot explain their existence and maintain current tenets of our science and theology at the same time. For example, what would happen if Sitchin's idea that the Israelite communication from the alien gods sometimes came through a radio were validated by the discovery of a receiver in the Ark of the Covenant? The psychological blow to the Judeo-Christian believer would be tremendous. To accept such intriguing individual artifacts and the presence of the ruins of large cities as evidence of advanced civilization, predating the current historical era, requires new psychological sets. But there is some reason to think such a widespread shift may be close. The above-mentioned 1996 television program *Mysterious Origins of Man*, hosted by no less a luminary than Charlton Heston and reviewing a number of the discoveries described in this chapter, found a receptive audience.

Sacsahuamán—a prehistoric center of wisdom and knowledge in Peru—is the site of hundreds of stone blocks, many weighing over 200 tons and placed more than two miles above sea level. How did they get there? Legend holds that Sacsahuamán was constructed hundreds of thousands—if not millions—of years ago, along with the Nasca Lines, the Pyramid of the Sun of Pachacamac, Lake Titicaca, the City of Ollantaytambo, the Sun Temple of Cuzco, and of course the renowned city of Machu Picchu. Thor Heyerdahl theorized that links existed between these sites and those of Egypt, Mesopotamia, and the Indus Valley.

In 1966, Robert J. Menzies discovered the remains of an ancient city off the coast of Peru where Inca ruins dot the shore. The city lies in the Milne-Edward Deep, a depression some 19,000 feet below sea level. Yet another example is the more-than-20,000-year-old Bolivian city, Tiahuanaco (pre-Incan "City of the Dead"), with artifacts of a highly developed civilization and portrayals of people very different in height and features from modern inhabitants.

These monuments to conscious achievement challenge our traditional accumulation of knowledge. The famous psychic Edgar Cayce reported visions of a civilization on Atlantis that existed for a period of 200,000 years, lasting until the last island sank beneath the waves around 10,000 years ago. We have always assumed that artifacts found in surface or near-surface sites were habitations from the first humans. After discovering the remains of beer in 5,000-year-old sites in Iran and Iraq, researchers concluded that the Sumerians were the first brewers. Perhaps they were continuing an industry that was already many millennia old, or even practiced in other worlds.

Did our human ancestors rise to heights of civilization yet unknown to us, destroy it, and then survive in remnants to start over again, and perhaps again? Or did other beings independent of the human family tree come and go, leaving their traces behind?

Both interpretations are worthy of investigation. Human beings may have existed throughout various ages, with a rhythmic rise and fall of scientific and social achievements. Cultures could have been destroyed or died from within, but small groups started again from nearly zero and then surged ahead. ETs could have introduced science and technology in places such as Sumeria, Nasca, and Central America. This external help would not contravene a multimillion-year local process of physical and mental development of conscious beings on Earth.

Missing Knowledge

The quality of the aforesaid unexplained artifacts and mysterious cities builds a strong case for the existence of intelligent life that predates conventional anthropological estimates. So why has their knowledge been lost? It seems almost as though it were wiped out at times and we had to start over again. Nevertheless, certain knowledge appears to have survived in some form.

All modern scientific discoveries may not spring fully blown from current research or by trial-and-error experimentation. Many believe advanced knowledge has been covertly held by a few in esoteric circles down through the ages. Another belief is that such knowing is latent in our collective memory. This book does not demonstrate proof of a particular explanatory theory of scientific insight or creativity, but it is possible within the theoretical framework presented that we are only now beginning to rediscover in the general field of consciousness concepts and techniques that were practiced on the planet long ago. Genetic engineering may be one of those areas of knowledge lost and rediscovered.

In the 1970s, Cabrera concluded from his analysis of the stones that genetic manipulation had been conducted on humanoid beings in prehistoric time. Interestingly, in early 1993 American physicians and researchers began seeking approval from the National Institutes of Health to administer such treatment to patients with brain tumors. The process involves removal of brain tumor cells which are then genetically altered and reinjected into the patient.*

Cabrera and Sitchin both believe extraterrestrials engaged in such genetic engineering to create beings to serve them, but some apocalyptic event caused the workers to be left behind by their designers. The early genetic experiments may have involved amphibians and reptiles before mammals. Experiments with the latter, in this view, led to modern humans. One theory is that the notharctus** (a false bear, hunter of insects and fruits), possibly the most intelligent of prehistoric animals (given its brain/body ratio), benefited from the insertion of cognitive codes more than 50 million years ago. Molecular compounds of nucleic acids and proteins could have been implanted at the embryo genetic phase by advanced beings. This theory implies that these genetically engineered beings (our ancestors) passed on to us the potential to reach the level of their manipulators or, with time and

---

* The theory is that these injections stimulate the person's immune system to fight the cancer. Some evidence from similar work in Japan has shown promising results.

** This could partially account for the legendary Arthurian origins of kingship.

experience, to go beyond it. Are we repeating on our own a practice imported by the "gods" thousands or millions of years ago?

Maurice Chatelain[18] believes that significant missing knowledge is not actually missing, only unrecognized. By taking into account the fact that the Earth's rotation is slowing at the rate of 0.000016 of a second each year, he inferred the Nineveh constant (the number 2,268 million that is a factor in many calculations of the solar system) was first calculated 64,800 years ago—a strong case for prehistoric intelligent activity. With additional mathematical analysis, he traces parallels between Egyptian and Mayan calendars, dating their origins to almost 50,000 years ago. Having found compatible instruments of measurement and similar bases of counting systems* in widely disparate ancient cultures, he is convinced of the existence of one or more unknown prehistoric civilizations.

Visitors or Immigrants

In addition to the presence of physical artifacts and intriguing theories of knowledge transmission, our collective cultural stories sampled earlier point to a multidimensional social history. Indeed, popular historical accounts (not to mention oral, unrecorded history), have recorded so much interaction with other intelligent beings that it may be a misnomer to think of them as "extraterrestrial." If they were commuters, we do not know if they were commuting to or from the Earth. Even if they originally came from elsewhere, such a long history of presence here places them more in the category of immigrants. Assuming that UFOs and similarly anomalous phenomena are external to our world may have led us up a scientific blind alley. They may be Solarians—even Earthlings—cohabiting this planetary system with access to advanced knowledge and other dimensions of reality. To better understand how we fit into the universe's schema, we need to understand how they fit in.

Since the flurry of UFO reports in the years immediately following World War II, thousands of credible sources have reported witnessing maneuvers

---

* The number 60 (squared at 3,600 and used as 360 for Earth-bound geometry) encompasses the methods of counting for the Sumerians (60), Mayans (20), Egyptians (10) and Gauls (12).

by alien craft. The first sighting covered in the national media was by Ken neth Arnold on June 24, 1947. In July 1947, occupied alien crafts apparently crashed near Roswell, New Mexico, offering tangible evidence of extraterrestrial visits. In attempting to discover the scientific level of Earthlings, the visitors may have pushed the limits of their technology too far, for other crashes reportedly occurred shortly thereafter.[19] Since then hundreds of sightings have been solidly documented by authentic photographs, traces left on the ground, and multiple witness descriptions. In the first third of 1992 alone, in Gulf Breeze, Florida, scores of appearances by anomalous craft (which continued into 1994) were photographed. On November 29, 1989, 125 detailed sightings of three-dimensional craft were recorded in a period of a few hours around a small town in Belgium. The national police and Belgian Air Force authenticated photographs and radar recordings of these craft and shared the information with the public.

In the United States, well-publicized alien contact began with the abduction of Betty and Barney Hill in 1961, was exemplified years later by the six-day abduction in 1975 of Travis Walton of Snow Flake, Arizona,[20] and continues until today. According to several Roper and Gallup polls, as many as two to five percent of the U.S. population believe they have experienced some sort of UFO contact.

To the question, why has NASA's SETI program (Search for Extraterrestrial Intelligence) failed to locate them? One answer is that we have been searching in the wrong place with the wrong instruments. Extraterrestrials are already here, operating in various folds of reality that we have arbitrarily excluded from scientific and conventional research.[21] Plowing a limited furrow, these programs are blind to the possibility of parallel rows to cultivate. Science is limited not only by narrow vision. Despite waves of documented appearances in most regions of the world at some time over the last 45 years,[22] the U.S. Government has kept such evidence highly classified, denying public access to its data. It is believed to include information, and possible physical evidence of crashes or debris from as far away as Norway and Brazil, and as near as a farmer's field close to Washington, D.C.

In this context, all reasonable approaches to discovery and contact have something to contribute to a more accurate understanding of who these beings are, where they come from, and what our relationship with them is. Some groups attempt to manipulate UFO behavior by simplistic signals, in hopes of precipitating a landing of an ET treaty delegation. However, history is already replete with many ET contacts and, from their perspective, beings who have been operating in our airspace, mind space, and home space for so long, the initiative rests with them. Given their superior mobility, technology, and scope of awareness, it is naive to think we can set the agenda.

An ancient Chinese tale describes small, gaunt men who descended from the clouds. For 40,000 years, Australian aborigines have passed from generation to generation a "dream" of space beings landing in their territory. (An aboriginal painting of that dream was hanging in the World Bank in Washington, D.C., at the time of this writing.) The sacred lore of Native Americans details the visits of ancient gods over thousands of years. The Mogollon culture in Arizona (located near the site of Travis Walton's abduction), along with those of the Navajos, Hopis, Pueblos, and Zunis, includes a very strong tradition of such visits. Similarly, in New Mexico, the Mescalero Apache Indians of Sierra Blanca tell of creatures from another world who came from the stars to interact with the ancestors of modern humankind.

Reports of ET contacts have had a degree of consistency over the centuries, even millennia, when superficial differences in cultural content are taken into account. For example, the abduction phenomenon of the late twentieth century smacks of the "incubus legends" of the fifteenth and sixteenth centuries. During that period, people in numerous small villages reported being visited by strange beings at night who sucked their vital energies from them. Similar waves of historical experiences include vampires, ghosts, poltergeists, etc.

Instead of dismissing all such traditional history as fantasy, scientists who would be serious must bring analytic minds to the study of past and current interactions with the apparent extraterrestrials. Some of society's

---

* The depictions of priests (Mayan and others) drinking blood from various artifacts may represent rituals modeled after ET blood transfusions and organ transplants, or other medical practices. It is possible that naive humans—in an attempt to align themselves with the so-called ET "gods"—later emulated the form without understanding the substance of ET practices.

present problems may conceivably stem from such external visitations, the impact of which could be negative or positive.* Could our violence toward each other derive from the example set by warring "gods" (as in Greek mythology)? Could human violence have been predisposed by physical or intellectual abuse? Would public confirmation of these experiences free humans from such long-forgotten influences, as the recovery of childhood trauma releases the dead hand of the past?

The beings alluded to above were so impressive that our ancestors called them "gods." As they returned to the skies, some of the "gods" may have given instructions pertaining to the care of the Earth and its beings, instructions which we took as holy writ. For some humans, the threat of the alien return was enough to stimulate dread and anxiety, which demagogues used to gain control over them. The aliens' lessons about inner powers could have been turned into worshipful rituals by less knowledgable humans, to be performed as rote without delving into their value. Failing to take seriously their access to cosmic powers, humans saw no need to heed the principles, as the Jewish people were explicitly exhorted to do by "visiting aliens" through the Old Testament book of Leviticus.

A new unabridged history, dealing with all the artifacts and legends from "the ages of the gods," would deflate our egotistical vision of ourselves as the first and most advanced sentient beings. A more accurate picture of Earth's past would include a history of cyclical rising and falling of civilizations, some with planet-wide impact; cataclysmic events—resulting from conscious actions or natural forces—that have destroyed civilizations and technologies; and intermittent engagements with extraplanetary groups, involving very high levels of consciousness. With this more accurate historical foundation, humans would be better able to understand our current institutions and future potential.

World War II marked a threshold of technological development that may have stimulated renewed extraterrestrial interest in Earth. Humans harnessed the conversion of mass to energy in a tremendously powerful nuclear weapon, perfected jet propulsion technology that presaged travel to the stars, and mounted a powerful radar system to reveal aerial surveillance. All these developments occurred in the late 1940s at U.S. Government installations in an area of New Mexico proximate to the locations where alien aircraft reportedly crashed in 1947. The last 50 years have seen a continuing escalation of UFO/ET activity. Only a small part of it is described in this book.

The moment may be approaching when, having completed their current reconnaissance or research objectives on humans and animals,[23] ETs desire formal contact. Perhaps for the first time in modern history, humans may be in a position for effective, two-way, mutually beneficial engagement. That will require the demise of the unilateral, and independent, government cover-up of knowledge and contacts that has involved every president since Harry Truman, and applies to senators and congressmen alike. Politicians who publicly admit UFO/ET experiences or interest, such as Carter, Reagan, and Goldwater, suffer the censorship of the covert system.

Unicosm as Home Base

Arthur C. Clarke (the science fiction writer whose work inspired the film *2001: A Space Odyssey*) has proposed that Earthlings might begin gardening the planet Mars. Drawing on recent research (by David A. Paige of UCLA) that shows water deposits frozen just under the Martian surface, Clarke postulates that we could cultivate grains and forests that would serve as catalysts for the formation of lakes and oceans. Clarke—who in 1946 predicted our current network of space satellite communications systems—sees our agro-development of Mars as a matter-of-fact extension of human development on Earth.

Responsible extraterrestrial exploration and stewardship are more than an extension of humanity's scientific and technological quest. They are a natural consequence of our cosmic consciousness. As we come to understand that our "divine" essence is not Earth-bound, we realize that neither is our physical presence. Our heritage and our future are not confined to this planet. As we have been seeded and/or fertilized by beings from elsewhere in the cosmos, our birthright includes a comparable mission. This planet is only home base; the material frontier for expansion of our civilization is ultimately as unlimited as the infinite boundaries of our consciousness.

Our recorded social history is just as spotty as our knowledge of the planet's geological and physical chronology. Every day we learn how much more we do not understand. For example, new dating of hominid and humanoid remains almost monthly pushes the antiquity of one species or the other farther and farther into the distant past. As a consequence, history, once recorded in very certain terms, has to be rewritten. Revisions on a much broader scale, in the light of newly discovered facts outside the field of archaeology, will be a monumental task.

Scientists and educators are challenged to update and expand every discipline in order to incorporate knowledge of the extended human past and its multilevel cosmic influences. Earthlings have only a dim memory of wisdom from that past, but a core of understanding about our true cosmic nature remains in esoteric teachings.

What some religious traditions consider "the fall of man" may have been only expulsion into "the cold cruel world of adulthood," where humans were forced to learn to be self-sufficient, without the direction of more advanced beings. Perhaps, as Plato reported, the gods left us to our own devices with a few insights or scientific secrets that would provide keys to later understanding. Tales abound of Great Teachers appearing under different guises around the planet. Stories of a Jesus-like being appearing in India, Central America, and even Glastonbury, support this concept of historical intervention. Humans now must face the reality of their own role as interventionists, and venture out from home base in all three realms— body, mind, and spirit.

It should be obvious by now that the human story is much more complex than the one described in our formal history. One has only to weave together strands of knowledge gathered from science, cultural anthropology, and interdisciplinary analyses of historical artifacts to discover we are fascinating beings with a very colorful past. Like recombined DNA, this variegated material gives us a radically different perspective on life in our planetary unicosm. Enough pieces now exist for *an entirely new historical framework to shape a new living myth for humanity* —The Solarian Legacy.

NOTES

1. "Science and the Citizen," *Scientific American*, Nov. 1995.
2. Sitchin, Zecharia. *The Twelfth Planet*: Book I of *The Earth Chronicles* (Stein and Day: New York, 1976).
3. Such fragments sporadically break free and traverse Earth's orbit in erratic routes. According to a scientist with the U.S. Geological Survey, about 110 such asteroids, ranging from just over one-half mile to 25 miles in diameter, pose a potential threat to the Earth, just as the Shoemaker-Levy 9 comet did to Jupiter in July 1994. Perhaps up to 4,000 other large asteroids and small comets (ice balls) also pose danger to the Earth. NASA and the Air Force are developing technology to track these objects, and many scientists (members of the International Astronomical Union) want the U.S. and other nations to develop ways to disable any that approach the globe. They suggest using lasers, small asteroids, or nuclear explosions to divert or destroy them.
4. Berlitz, Charles. *Atlantis*: *The Eighth Continent* (Ballantine Books: New York, 1985).
5. Cabrera Darquea, Javier. *The Message of the Engraved Stones of Ica* (Servicio Grafico '2000': Lima, Peru, 1989).
6. Thompson, Richard, and Michael Cremo. *Forbidden Archaelogy: Hidden History of the Human Race* (Torchlight Publishing: Badger, California, 1996).
7. Milton, Richard. *The Facts of Life: Shattering the Myth of Darwinism*. (Corgi Books, Transworld Publishers: London, 1995).
8. Sproul, Barbara C. *Primal Myths* (Harper San Francisco: San Francisco, 1991).
9. Clark, Jerome. (1) *The Emergence of a Phenomenon: UFOs from the Beginning Through 1959*. (2) *High Strangeness: UFOs from 1960 Through 1979*. (3) *UFOs in the 1980's*. (Center for UFO Studies: Chicago, 1990-96).
10. Sprinkle, R.L. "The Changing Message of UFO Activity." *Proceedings*. (Conference of the International Association for New Science: Ft. Collins, Colorado, 1988).
11. Hancock, Graham. *Fingerprints of the Gods* (Crown Publishers: New York, 1995).
12. Bramley, William. *Gods of Eden* (Avon Books: New York, 1993).
13. Horn, Arthur D. *Humanity's Extraterrestrial Origins* (A & L Horn: Mt. Shasta, California, 1994).
14. Steiner, Rudolf. *Cosmic Memory* (Harper & Row: New York, 1959).
15. Hoagland, Richard C. *The Monuments of Mars: A City on the Edge of Forever* (North Atlantic Books: Berkeley, California, 1987).
16. McDaniel, Stanley. *The McDaniel Report: On the Failure of Executive, Congressional, and Scientific Responsibility in Investigating Possible Evidence of Artificial Structures on the Surface of Mars and in Setting Mission Priorities for NASA's Mars Exploration Program* (North Atlantic Books: Berkeley, California, 1994).
17. Velikovsky, Immanuel. *Worlds in Collision* (Doubleday and Company: Garden

City, New York, 1950); Pauwels, Louis, and Jacques Bergier. *Eternal Man* (Mayflower Books Ltd.: Frogmore, St. Albens, UK, 1973); and *Mysterious Origins of Man*. Narr. Charlton Heston. Prod. B.C. Video, Inc., P.O. Box 97, Shelbourne, Vermont 05482.

18. Chatelain, Maurice. *Our Cosmic Ancestors* (Temple Golden Publications: Sedona, Arizona, 1988).
19. Randle, Kevin D., and Donald R. Schmitt. *UFO Crash at Roswell* (Avon Books: New York, 1991).
20. The film *Fire in the Sky*, released in March 1993, details Walton's abduction in November 1975, when six friends saw him knocked down by a beam of light from a spacecraft. His book with the same title published in 1996 gives additional facts.
21. Vallee, Jacques. *Confrontations* (Random House: New York, 1990).
22. Hall, Richard. *Uninvited Guests* (Aurora Press: Santa Fe, New Mexico, 1992).
23. Howe, Linda M. *Alien Harvest: Further Evidence Linking Animal Mutilations and Human Abduction to Alien Life Forms* (L.M. Howe Productions: Philadelphia, Pennsylvania, 1989). With regard to animals, the aliens have taken the same callous approach as humans have—if indeed they are resposible for disssecting and removing organs from livestock in many areas of the world. Carcasses left behind contain the marks of advanced surgical procedures.

PART II

Expressions of Cosmic Consciousness

HUMAN BEINGS NEED A SENSE of the structure and dynamics involved in the expression of cosmic consciousness in the universe if they are to grasp fully their role and potential. Part II provides a three-faceted model for understanding the relationship of general and local consciousness to energy and matter. It reveals the roles of causative mind and subtle energies in the shaping of the material realm, including the power of individual beings. A hypothesis for incarnation is presented and developed in the model. The role of a self-defining self in maintaining the integrity of personality is explored.

Chapter 4

Structure: Three Facets

Reality, from the conventional scientific perspective, is unidimensional: everything is part of and limited to the material realm. Many philosophers think of the universe in dualistic terms: mind and matter. Some religions see three parts: body, mind, and spirit. Frontier science and metaphysics offer another concept: a multifaceted universe with all elements interactive and conterminous. In a three-faceted universe where consciousness is supreme, at least three types of senses are operative in conscious beings. Some laws of the universe are discovered by the physical senses, while others require the subtle senses. Scientists need a more expansive framework if they are to offer comprehensive answers.

◆ ◆ ◆

HAVING EXPLORED the macrocosmic and microcosmic dimensions of the universe and the unicosm (planetary home base) that holographically comprises the full spectrum of its nature, we now turn to the modes through which conscious beings relate to their universe. For classical Greek philosophers, the concept "phenomena" included all concrete things in objective reality. But they also believed in a nonphenomenal realm, the permanent essence of things which could be known through intuition. Known as the "noumena," this was the subjective domain of ideas and abstract forms. The two are bound together by a third realm of subtle energy or forces labeled "energeia" in Chapter 2. The universe is composed of at least these three interdependent realms, or facets of one whole, that are accessible by humans.

The noumenal and energeial, or intuitive, ways of knowing have been devalued over the last four centuries by the growing global acceptance of the European-based, Newtonian view of physical science. The result is that currently all formal human institutions of significance—those defining our "official truth"—are dominated by phenomena-based thinking.[1] Most people believe that being objective in any aspect of life, and particularly when undertaking scientific research, means limiting oneself to five channels: smell, touch, taste, hearing, and sight. But not everyone so limits themselves.

The Bushmen of southern Africa described in the books of Laurens van der Post have not given up their sense of "tapping," a sensation deep in the chest which indicates they are on the track to food or relating properly to their habitat or community. The water-well or lost-object dowser continues to develop a feel for a presence that cannot be seen or touched. Serious scientists, like an Einstein or a Tesla, do not reject solutions that come in visions or dreams. The clairvoyant does not stop using an inner eye. The successful businessperson does not ignore hunches. A fully conscious person combines all these ways of knowing.

Design—Forces—Matter

Although modern science has limited itself to verification by the physical senses, refined instrumentation, extending the reach of scientists' sight and touch, enables them to manipulate the subatomic realm of ordinary matter. But this success has its price. Stuck on the tactic of trying to understand the inner workings of materiality by crashing through its outer walls, researchers use physical hooks to open the doors to atoms, molecules, cells, organs, and bodies of all kinds. They observe events like chemical reactions, tumors, comets, storms, or flowers, and immediately start looking for the physical causes. This approach only leads to other material connections which, as we have seen in the microcosm, become more and more difficult to pin down.

To understand mind and emotion, the mysteries of conscious life, we must conceptually reach beyond the *phenomena* to infer the dynamics of various forces at play *(energeia)* and to infer the patterns and principles behind those dynamics *(noumena)*. Making such inferences does not lead to a theoretical cul-de-sac: their implications can be tested in the mundane world, including areas of human experience now outside the explanatory power of traditional science such as mind/body healing and telepathic communications.

A three-faceted model is an appropriate concept, with components analogous to what Gregory Bateson might have labeled "substance" *(phenomena)*, "energy" *(energeia)*, and "no thing" *(noumena)*. Michael Talbot would have labeled them, respectively, "mass," "energy," and "information."[2] Parallel Christian concepts (Father, Son, and Holy Spirit) and teachings of the Jewish Kabalah point to the same insights in esoteric knowledge. A new renaissance scientist will recognize the model in the Kabalah's three levels: Nefesh (state of rest), Ruach (wind), and Neshama (breath of god). In the Kabalistic view the "breath of god" (logos or *noumena*) activates matter through the "wind" *(energeia)*. The ancient Hindus used AUM with each letter representing one of the aspects.

Use of these metaphysics helps create a broader science: a *metascience* that underpins a more comprehensive approach to cosmic life. They point to theories that postulate an intermediate or subtle energy field with its own set of forces responsive to permutations of thought or ideas. They enhance our understanding of the processes that, to use the concepts of quantum mechanics, collapse a potential pattern of reality into experienced or observed reality.[3]

The *phenomena* and the *energeia* exist as potential to be activated by conscious intent through the expressive and receptive modes of the physical and subtle senses. In the discussion of the void in Chapter 2, we saw conscious intent has an indirect impact on energy and matter. To grasp how that intent is channeled and manifested involves an understanding of subtle energy forms and their kinetic force.

Scientific knowledge of all energy spectra is still fragmentary, with our understanding becoming more diffuse as we move from left to right on the continuum shown below. We understand gravity in the mechanical arena and chemical energy expressed as heat in the conversion of matter, but the other areas grow more fuzzy as one moves to the right. For example, Robert Jahn of Princeton University believes consciousness interacts with the material realm through the quantum wave function, but where does that interaction fit on the continuum?

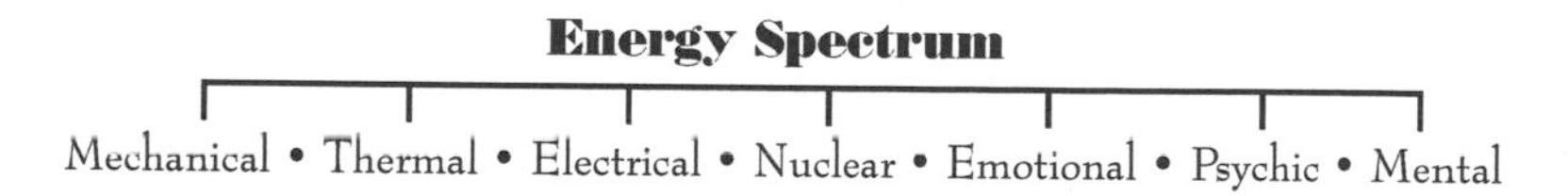

The common applications of energy apply to the phenomenal realm (electrical, nuclear, solar, muscular, etc.). New concepts of energy beyond the electromagnetic spectrum are appearing. Research with scalars—waves that continue to reverberate whenever matter is removed from space—is leading to progress in the development of so-called "free energy" technology. Scalars have been considered cosmic waves by some, because they are believed to be faster than the speed of light, leaving no trace in the material realm. The term "zero-point energy" (ZPE) is used by some to refer to this still-mysterious energy. This text employs the term "subtle energy" when referring to energy in the energeial realm to avoid confusion with ordinary energy that is the obverse of matter.

Waves like the above that are not part of the electromagnetic spectrum are the subtle energies of the *energeia*, which, in the model offered here, arises from consciousness, as *matenergy* appears to rise from the *energeia*. We need new hypotheses to help test how an idea acts on subtle energy that in turn shapes *matenergy*. (Recall that *matenergy* denotes a combination of matter and energy as two points on the same spectrum in the phenomenal realm.) Research projects should begin with the assumption that energy in the phenomenal and energeial, and probably the noumenal, realms possesses both potential and kinetic states. Interestingly, the ancient Tao philosophy expressed in the Tai-Chi Chuan school of Chinese martial arts says Wu-Chi, the state of nothingness, holds the potential for both static and dynamic states.[4]

In its potential form, physical energy is force waiting to be released, whether a precariously balanced rock, a piece of wood or food, or an atom. In the three-faceted model, embodied (incarnated) emotionality is derived from potential subtle energy; and local mind includes potential thought patterns awaiting unfoldment. In our four-dimensional world, kinetic energy involves movement through space-time, be it in the rebounding billiard ball or a surge of electrons. In the movement of air molecules, the flitting of electrons, and the pulsing of prayers, the isomorphic principle is that transformation of energy is an *expression of intent*. At all levels, energy transformation from potential to kinetic is the *expressive aspect of some form of communication*.

When we understand that various forms of energy and matter come from nonmaterial events, we open the doors—through the use of *all* our senses—to greater expression of our inner power. The making of imaginal choices creates thought patterns that serve as energizers. Between ideas and emotions lies the interface where the subtle energies of the emotional realm *(energeia)* are activated by the mind's patterns. These emotions in turn, through the media of electromagnetic, mechanical, and chemical energies, affect the phenomenal realm. The earlier discussion of the role of ideas and emotions in the creation of neuropeptides offers an illustration of this sequence. The fully conscious being in the universe requires multilevel senses to act as gateways between these three realms. The following review of different categories of senses illuminates the dynamics of the three realms in the context of a conscious universe.

Five Physical Senses

In her book *A Natural History of the Senses*,[5] Diane Ackerman uses the written word in a sensory exploration of a world teeming with physical stimuli that define the limits of consciousness. Her beings luxuriate in a flood of wonders that, while sharpening their focus, actually limits their experience. Given the quality of interaction available within the rich material environment, it is no wonder that many of us end up stopping there.

The five ordinary senses are the entry points through which data from the denser material plane connects with the self. At the outer perimeter of the physical body, these specialized cells receive input from the external environment and send it to the brain for analysis. It is helpful to think of the sensory receptors as corporeal extensions of the central nervous system, analogous to instrumental extensions like the telephone, the telescope, the stethoscope, etc.

In each category, a sense organ does not make sense of anything; it only takes in bits of data. The mind, working through the brain and the nervous system, makes sense of what is perceived. The most appropriate terms for the five senses, therefore, are *receptive* ones. The eyes *see,* rather than look. The skin sensors *feel,* rather than touch. These examples make it clear we need to be careful with our terminology. Although receptive, the five ordinary senses are not passive. For this reason, the gerund (noun) form of verbs—seeing, hearing, smelling, feeling, and tasting—are used here.

Seeing

Seeing is our perimeter guard post for understanding what is going on in the world around us. Its range is the longest of the five ordinary senses, literally to stars deep in the universe. Seventy percent of our body's receptors are devoted to it. The retina takes in the light waves and focuses them on photosensitive rod- and cone-shaped cells. Different cells, perceiving different colors on the electromagnetic spectrum, send electrochemical signals to the brain's visual cortex. The visual cortex—used by the mind to make sense of these signals—does the actual looking. (This is important to remember later when we discuss televiewing.)

The visible light our eyes perceive constitutes a very limited range of the electromagnetic spectrum. The old saying "what you see is what you get" is misleading. We get much more than we see: the richness of all light helps us survive and thrive in our habitat. The light not seen by our eyes affects biological as well as psychological rhythms. Our development and health are bound up in the cycles of light and dark.

Hearing

Hearing is the registering of sound, which is simply ongoing waves of air molecules. These moving molecules (like the circles caused by dropping a stone in a pond) can come from the vibrations of any object. Like waves on the beach, the sound waves crash on our eardrums, making the eardrum vibrate, tripping the tiny hammer, anvil, and stirrup-shaped bones. These bones push fluid in the inner ear against membranes that jostle tiny hairs, which in turn cause nerve cells to send electrical impulses to the brain.

Sound waves, like all others, have certain frequencies. High-pitched sounds have high frequencies. We can generally hear frequencies between 16-20,000 cycles per second, a range of about ten octaves. Vibration over 20,000 cycles per second is considered ultrasound. (Although ultrasound cannot be heard, its force can be harnessed for a variety of tasks in healing, heating, lifting, etc.) And though the ear is particularly fine tuned, our entire body registers sound waves—acting as a huge acoustical receiver.

Smelling

In our noses are olfactory sites—the neurons in five million receptor cells—that are sensitive to odorous molecules drawn into our system. This drawing in is a receptive process, an integral part of our inhaling. Upon exhaling we send the molecules back out, our brain's limbic system having interpreted their electrical messages about the state of the contiguous world. Once again we have the yin and yang: a person both smells (senses) and gives off smells.

Smelling is one of the body's front-line defense systems. Through smell the hypothalamus senses nearby danger, mating potential, food sources, and other factors related to survival (like toxic air). Odorous changes in the earth and in people, through the medium of pheromones, stimulate biological reactions—adrenaline rushes, salivating reflexes, and sexual changes (including shorter menstrual cycles). Artificial pheromones can be used to stimulate affectionate or rejecting response patterns.

Feeling

The functions of the skin—a two-layered membrane less than two millimeters thick—are to feel and to serve as our physical boundary. The largest organ in the body (six to ten pounds), the skin keeps more in than it lets out, but is by no means impermeable: its cells breathe and excrete as all cells do. The skin blocks out some of the sun's rays, fends off microbes, serves as an insulator, and metabolizes vitamin D. The outer layer, the epidermis, constantly renews itself.

The inner layer of skin—the dermis—has several kinds of sense receptors that respond to pressure. More receptors are concentrated in hairy areas of the skin, hair itself being a receptor. These receptor cells, depending on the quality of contact (heat, pressure, or vibration), respond with a repertoire of electrical signals to the brain. Crucial for psychological as well as physical well-being, feeling is the first sense that becomes operative for the newborn.

Tasting

Tasting, like smelling, involves ingestion, but it is a much grosser screen:. our taste buds require thousands of times the number of molecules required for smell. But, like smelling, tasting also requires that geometric molecules fit the receptors in order for electrical impulses to be sent to the brain.

Tasting is our ultimate line of defense against threatening substances that might otherwise be taken into the body. It also helps us know what is good for us. Taste is a measure of likes and dislikes: to have a taste for something is to need it or to desire it. There is a direct relationship between our tastes and our subjective states.

Senses in Context

All five physical senses are interrelated, working in tandem to keep us informed of our situation relative to other beings and our environment. For some people, like my daughter, who says colors have smells, one sense stimulates another. In the extreme, this condition is called *synesthesia*. Although no single sense is absolutely essential for life, together they make navigating our physical environment easier and more pleasurable.

The five physical senses seem to deal with such different kinds of information that we unthinkingly consider them as separate and distinct. Once viewed in the context of the electromagnetic spectrum that shows their essential similarities, the perceived "gaps" between these senses disappear. Whether light waves, sound waves, or pressure waves, each and all are perceived by specialized organs of the same integral human body that uses them to orient itself in the pool of universal *matenergy* in which it thrives. Whether the chemical fit is sensed by the nose or the tongue, whether the pressure is sensed by the skin or the ear, the result is the same: a sample of the immediately contiguous physical environment.

Each of the five physical senses perceives only a part of the range of stimuli for which it is designed. We see only certain light waves, hear only

Table VII

ELECTROMAGNETIC SPECTRUM AND HUMAN SENSES

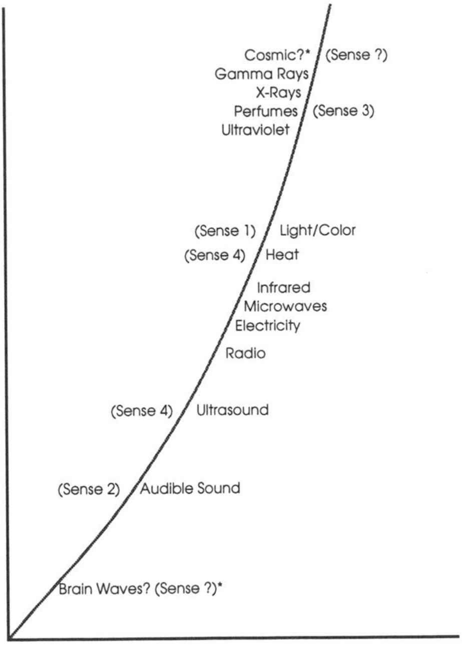

Cosmic?* (Sense ?)
Gamma Rays
X-Rays
Perfumes (Sense 3)
Ultraviolet
(Sense 1) Light/Color
(Sense 4) Heat
Infrared
Microwaves
Electricity
Radio
(Sense 4) Ultrasound
(Sense 2) Audible Sound
Brain Waves? (Sense ?)*

Frequencies (Not to scale)

* Are these different points on EM spectrum or different spectra?

limited frequencies, and smell only certain scents, etc. Table VII reveals how much we miss on the electromagnetic spectrum. The range of stimuli we perceive in each sense spectrum seems to be related to our immediate survival needs. Consider how dull cosmic existence would be if we were limited to just five senses. Fortunately, we have a larger set of capacities at our disposal.

All senses always involve the expression (masculine) of the input which is perceived (feminine)—the yang and the yin. There must first be an expression toward us that we then sense; conversely, without an ear to hear, the falling tree makes no noise. The connections delineated in Table VIII indicate the interplay between expressing and sensing. We frequently confuse *sensing* with *expressing* by using the same word for both the yin and yang acts. We say we are touching something external to us when we actually are feeling the touch of someone or something else. When something has an odor, it has a "smell" and we "smell" it.* When we hear, we are detecting the vibrations of something else. The ear cannot hear its own vibrations. The same is true for the subtle senses.

Humans have at least five subtle senses that parallel the five primary physical ones. In fact, the Principle of Correspondence would indicate that all physical senses have analogous subtle senses in the emotional and mental planes.** (See Table VIII.) All specialized channels through which we interact with each other and the universe are in effect senses, so we are likely to have fifteen or more identifiable senses. If the seven main chakras have related subtle energy senses—as I believe they do—then the number of identifiable human senses is even greater. These subtle senses are our natural gifts: their full use can arise either spontaneously or through conscious effort.

Examples of animal senses outside the ordinary five can help us grasp the idea of the subtle senses. Some animals are extremely sensitive to one or more physical stimuli: magnetism,[6] electrical fields, ultraviolet light,

* Mark Twain allegedly remarked to a haughty lady, "Pardon me madame, but I stink and you smell."

** I tentatively label a third sensory level, beyond the subtle senses described here, as the senses of consciousness (sc), but do not attempt to further develop an explanatory theory. Further exploration of this level is needed to move beyond the current tendency to lump several noumenal senses under terms like "clairvoyance" or "psychic vision."

polarized light, barometric pressure changes, and infrasounds. Animals can pick up these perturbations from great distances, sometimes hundreds or thousands of miles away. Some birds may guide themselves on long flights by matching imprinted patterns with sun angles and constellations. Bees, too, use polarized light from the heavens to navigate. Bats fly sensing sonic reverberations much as jets use radar. Fish respond to changes in the salinity gradient and "hear" through the vibrations of small stones (otholiths) nestled in a bed of nerve hairs on each side of the brain. Sharks are attracted by small increases (just a few parts per billion) in the blood gradient in water.

Humans may have residual forms of some of the animal senses mentioned above. Some of our seemingly disoriented behavior may be caused by perturbations in senses we do not recognize we have. We also appear to share more subtle senses with animals and/or more evolved beings.

Table VIII

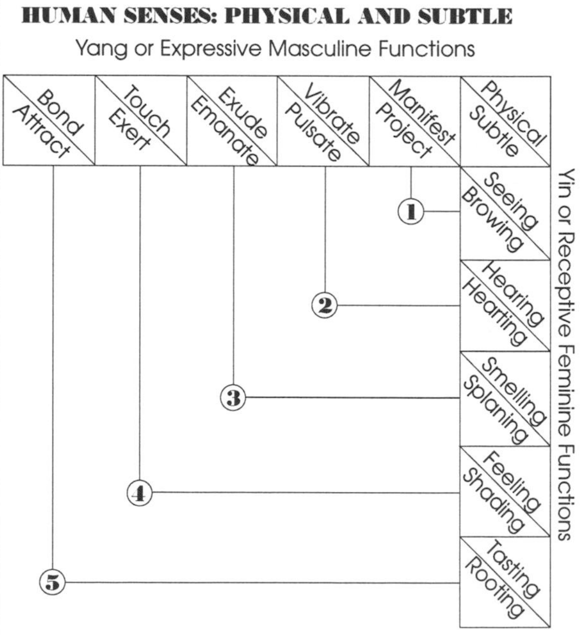

HUMAN SENSES: PHYSICAL AND SUBTLE

Yang or Expressive Masculine Functions

Bond / Attract	Touch / Exert	Exude / Emanate	Vibrate / Pulsate	Manifest / Project	Physical / Subtle
				1	Seeing / Browing
			2		Hearing / Hearting
		3			Smelling / Splaning
	4				Feeling / Shading
5					Tasting / Rooting

Yin or Receptive Feminine Functions

Note: The "senses of consciousness" are another level on this matrix.

If the five senses based in the phenomenal realm reflect the characteristics of that realm, the subtle senses help distinguish the differences between the realms. For example, the subtle senses would not be limited by time and space, nor would they be dependent on *matenergy*. In effect, we can infer the realm or dimension from which any sense operates by its characteristics. The "tapping" of the Bushmen works through the "*energeia*," while precognition of images originates in the "*noumena*."

Five Subtle Senses

As the physical senses are all characterized by their electrochemical nature, the senses of the *energeia* involve the subtle energies of the being's emotional field, and those of the *noumena* convey the power of pure thought.* The five senses described below are, like the physical senses, receptive and interrelated. They are reciprocals of the complementary expressions of subtle energies in currently unknown spectra parallel to the electromagnetic spectrum.

Browing

"Browing" relates to the ordinary sense of seeing. A form of televiewing, it is not limited to line of sight or the speed of light. It is our channel for sensing coherent forms of subtle energy. People who can "see" auras understand this sense without difficulty. But all of us can perceive images within the *energeia* without regard to physical location, both in local time (remote viewing) and beyond time (precognition of subtle forms and retrovision).

With "browing" we can access the energized forms that derive from strong conscious activity as well as the ongoing energy fields of phenomenal events or artifacts. Such thought forms may also include energized group perceptions of events in the general storehouse of memories. The

* I suggest labels for five senses of the "*energeia*," recognizing that it is only a beginning and that further conceptual development and research are required. My taking the liberty of creating new terms is an effort to bring a fresh perspective to the senses that are often indistinguishable under labels like "intuition," or "sixth sense."

events themselves are only transitory and are not stored; only the memories are. In the Hindu chakra system, "browing" relates to the third eye or sixth chakra.*

Hearting

To "heart" is to monitor shifts in frequency that result from changes in the love/hate (attraction/rejection) polarity of another being or group. Relating to the fourth or heart chakra, the vibrations of this subtle force and those of sound are analogous: "Hearting" and hearing have the same root syllable. In both realms, these parallel forms of energy can be either highly strung and defensive or serene and malleable, depending on their frequency. Making behavioral waves is therefore a visible manifestation of the subtle negative energy vibrations.

With this sense, our interpersonal disposition can be perceived by those attuned to us, even over vast distances. Through this sense one "hears" the cry of one seeking to be alone or "knows" of another's remote reaching out. When we tell another, "I not only heard you, but I really hear what you're saying," we are attempting to reflect both the physical and subtle dimensions of the capacity to communicate.

Splaning

Through our ability to "splan" we sense another's emotional state or humor. One's general degree of openness versus defensiveness is communicated through the spleen area, or the site of the second chakra. This sense measures the expression of the fight/flight polarity: we "splan" projected incursions by others into our Selfhood or, obversely, withdrawals from the potential engagement.

On a subtle level, it is almost as if we can smell the intentions of another being, not unlike an olfactory engagement with a skunk in bad humor or the hormonal signals exchanged between potential sexual partners. Through

---

* "Chakra" means wheel in Sanskrit[7] and comes from the wheel-like vortices that exist in the subtle energy field of a human being. Chakras serve as channels for the flow of this life force into the physical body.

the sense of "splaning," one is aware of the unspoken desire of the other regarding degree and quality of openness.

Shading

The "shading" sense helps us register the state of another's aggression index, or the degree of expressed nonphysical pressure. (Some expressions of telekinesis can be perceived through this sense.) A lion signals its impulse to dominate territory or exert control in a manner that is obvious to the physical senses; humans send out similar signals, but there are others not necessarily so obvious. Beyond overt behavior and the field of chemical and electrical exchanges, such proactive impulses are also communicated through the medium of energetic patterns at the prematerial level, perceptible through "shading."

Received at a distance beyond the range of the physical senses, through the third chakra (in the navel or umbilical area), this sense registers the umbra, or shadow, of another. When attempting to describe our reactions to data received through this sense, we often say we are having a gut or visceral reaction; it is almost as if we can feel the invisible shadow cast in our direction.

Rooting

Our "rooting" ability serves the same role for individuals as the court taster does for the king: It keeps us alive. Energetic communications warn us of an impending noxious encounter or seduce us into agreeable interaction with other cosmic beings. The basic inclination to joint endeavor is perceived through our first chakra, even without our being face-to-face.

Messages transmitted through this sense of energetic compatibility (mistakenly called "chemistry") are much more reliable indicators of interpersonal harmony than common physical characteristics. The survival of individuals, families, and races depends on the reliability of this channel of communication. In such important matters, the messages exchanged through the root must take precedence over those of the heart or the spleen. The energetic "taste" at this long-wave level is a truer indicator of sustainability than the higher frequencies of sound and vibration received by the ear and the heart.

In the subtle realm, as in the physical one, when we receive the vibration of another, focused awareness is necessary to make sense of (and validate) the subtle communication. The value we consciously attach to energetic signals determines the ultimate nature of our reaction, as in the tasting of food. Etymology, the study of language development, indicates that the sense of color differentiation has become more complex in humans over the centuries. We can now describe more differences in the wave lengths and amplitudes (shades of color) on the light spectrum than we could previously. We do not know if that is a result of evolution in the physical eye or in the mind that interprets its input. Regardless, the same process will undoubtedly refine the capabilities of our subtle senses. The first step is to become more consciously aware of them.

Senses and Consciousness

Neural networks that send and receive expressions are involved with all the physical senses. They are now being simulated in the mechanical world of computers. Although subtle senses can be categorized as remote sensing—there are no physical *(matenergy)* connections—some medium appears to be involved in this so-called paranormal activity. Cleve Backster has found that cellular reactions *precede* transmission of material or energetic stimuli in the nervous system, indicating the *non-time-consuming* nature of mind transmissions. The corresponding nature of sensing within the phenomenal and energeial realms may imply that the senses are the channels of reciprocity between them. The sense nodes of our being may be where the overlap between realms occurs, and, therefore, the channels of creation. For example, the subtle senses may be fundamental to incarnation, perhaps to the point of serving as channels for transmitting the energeial precursors to DNA patterns. Thus preexisting conscious intent works through the subtle medium to manifest itself in matter.

The subtle senses stand as sentries at the perimeters of the emotional body, as the physical senses do for the physical body, thereby serving as the interface between materiality and emotionality. The analogous senses of consciousness are involved in conveying patterns or ideas from the *noumena* to the *energeia*, the field of subtle energy. Through these connections, consciousness has the ability to transform subtle energy.

All cosmic beings are comprised of various manifestations of the three facets, in effect, having three overlapping bodies. The emotional body is subtle energy infused with consciousness. The physical body is materiality infused with emotions. The mental body is a local manifestation of the universal mind. The three so-called bodies are not separate, but are aspects of the same being perceived from distinct but interrelated perspectives. The following diagram is an attempt to illustrate these aspects and the senses that facilitate their interrelationships. Although a two-dimensional illustration does not adequately portray the conterminous nature of the three bodies, it indicates something of the various interactions, internally and with the environment.

Cosmic Beings

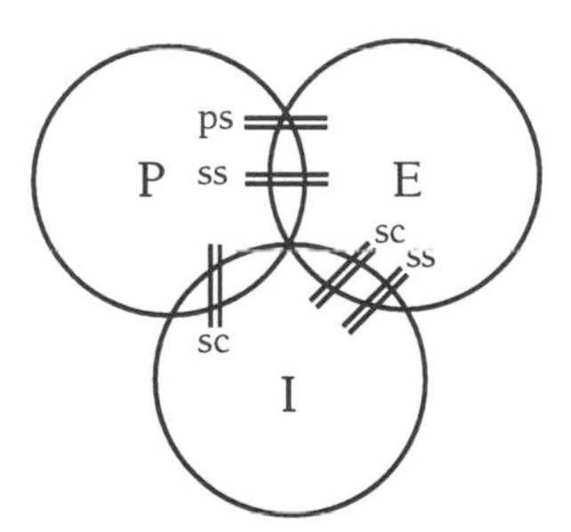

Key

P = Physical (Phenomenal) Body
E = Emotional (Energcial) Body
I = Imaginal (Noumenal) Body
ps = physical senses
ss = subtle senses
sc = senses of consciousness

According to this schema, when a physical event is perceived by the physical senses (ps), the stimuli are converted to electrical impulses that trigger physical responses. Simultaneously, the event's subtle stimuli are perceived by the subtle senses (ss), where they are channeled, via the as-yet-undefined senses of consciousness (sc) to the individual mind for interpretation. (This is why a physical act done with loving intent is received differently when the same act is undertaken with hate or anger.) Conversely, when consciousness creates an idea, it is conveyed via "ss" to the emotional body, where its intent and force is communicated via subtle energy to the brain/physical body, whereupon the body is activated. Thus the singular and integral nature of the three-faceted being is maintained.

The electrical impulses from the physical senses that stimulate reactions in the brain and its nerve system have been mapped. The subtle senses involve a similar or corresponding process of passing impulses through the emotional field, but we have no means currently to identify it. (The work done by chakra and acupuncture specialists may be way ahead of Western scientists in this regard.) Even more elusive is the process through which thoughts energize the emotional body. Although the local mind can observe directly the physical result of its idea, it can only reshape it through the medium of subtle energy. Directions of flow in the above diagram provide for this hypothesis.

If the emotional body (coalescing of subtle energy) and imaginal body (focusing of the idea) are the intermediary stages between the material dimension and the universal *noumena,* then many previously unexplained "paranormal" events become fathomable and testable within this model. Any subtle event (remote viewing, psychokinesis, healing touch, answered prayer, etc.) should parallel a physical one and be connected through the senses. With these basic insights, one can extrapolate descriptive and explanatory theories for even the most bizarre unexplained events.

In India, for example, thousands have witnessed Sai Baba and Baba Altamas materialize objects that can be seen and felt like any other phenomena. Some yogis believe the vibrations of such acts (see the expressive functions in Table VIII) are controlled by the life force they call "prana," which in turn is made up of "lifetrons."* They believe the regulation of this pranic force by master practitioners permits them to rearrange the vibratory structure of the inert void to materialize perfume, fruit, and other physical objects. A kind of reverse engineering, taking effects on the human body as objects for research in consciousness, is required to uncover this power of mind over matter. After all, our bodies are excellent examples of the application of intentionality: they are the manifested desires of a certain—we do yet not know what—quality of consciousness.

* All sophisticated traditional cultures had a similar concept. For the Mayans, it was known as "Ch'ulel."

Filling the Gaps

David Hopes wrote, "Having discovered the wrong laws does not entitle us to declare that the universe is lawless."[8] Just because we cannot fit our experience into conventional laws of science does not mean other laws do not exist. What we have called "being realistic" is actually a censoring of the parts of reality we are not willing to examine, or are not yet intelligent enough to understand. Human fulfillment of its potential role requires understandings that go beyond conventional scientific explanations of many phenomena, including filling in gaps like the following:

- UFO/ET experiences
- free energy sources
- healing arts
- psychic phenomena

The following examples of data from unexplained human experiences indicate realms of activity, energy, and knowledge that would enhance human performance, offering potential solutions to some of civilization's most pressing problems, if we were to tap them. They deserve scientists' priority attention.

Ufology

That ufology is not regarded as an "official" field for investigation illustrates misguided, selective inattention by most of the scientific community. The result is that many hypotheses addressing the origins and nature of UFOs are not well thought out, lacking repeatable and verifiable tests. Confusion also exists about which reported sightings and contacts are actually those of UFOs, in contrast to those that can be attributed to ordinary events. Nevertheless there are literally thousands of such reports which are not neatly rationalized. Only people who are afraid (for personal or professional reasons), or totally given to prejudgment, can ignore the mounting evidence. Although untold thousands of reported experiences with the occupants of UFOs (generally assumed to be ETs) have been persuasively documented, we still have no generally accepted explanations for such beings and our relationship to them.[9] Regardless of whether the occupants of UFOs intend to reveal themselves more dramatically any time soon, the knowledge base of the general public would be considerably enhanced by serious, systematic study.

Since the 1950s, UFO fragments have reportedly been recovered in several countries. For example, a piece of debris from a reported UFO explosion over the sea near Ubatuba, Brazil, was turned over to the Brazilian Mineral Products Laboratory. The fragment was subjected to neutron activation analysis and revealed to be almost pure magnesium, with only .001 percent impurities (barium, zinc, and strontium). This tiny degree of imperfection in the crystalline structure of magnesium is beyond any current state-of-the-art production capability. In the U.S., pieces alleged to have been from the Roswell crash and others have been subjected to similar analyses with equally mysterious results. The U.S. Government's Operation Moondust has sought such items around the world for study.

The electronic equipment of airplanes, police vans, and civilian cars has been affected by UFOs in their vicinity. Magnetic traces have been left on the vehicles. Gauss meters at UFO sites have shown unexplained variations in magnetic fields. Similar anomalies in gravity measures, as well as physical traces, have also been reported.[10] Many persons have had various physical reactions—rashes, bruise marks, and lesions—to UFO radiation. Widespread incidence of animal mutilations, frequently involving the reproductive organs, have been extensively documented by Linda Moulton Howe[11] and others. Humanity deserves expanded scientific studies of these unexplained mysteries in the physical sciences.

Free Energy

Another area deserving open-minded exploration is that of alternative energy sources that hover at the edges of the phenomenal world. Bruce DePalma, Troy Reed, and others developed one such alternative source—sometimes referred to as "free energy"—by working directly with the shape of magnetic fields. Instead of switching magnetic fields off and on as we do with conventional electric motors, Reed's "magnetic motor" produces mechanical energy (turning a drive shaft) by manipulating the shape of magnetic fields. DePalma's related work with zero-point energy is being further developed by researchers in India and Japan.[12]

Early claims about successful experiments with cold fusion, another "free energy" hypothesis, were initially dismissed by the scientific community and the popular media. Now scientists around the world are validating the 1989 findings of Martin Fleischman and Stanley Pons, whose work (in France) is continuing with support from Japan (whose lack of fossil fuel deposits gives it more reason to be open to this new science).[13]

While the "free energy" concept appears to violate the second law of thermodynamics, it may imply that we are expanding our understanding of the scope of physics. Harold Puthoff[14] and other physicists have demonstrated in laboratories the existence of energy in a vacuum that is more powerful than a nuclear force and that can be tapped without destructive side effects. Even though the International Association of New Science and its Institute for New Energy held four excellent conferences on free energy from 1993 to 1996 in Denver, Colorado, little mainstream media attention has been given to these scientific breakthroughs.

Healing Arts

Many areas in the healing arts await serious and widespread scientific testing and validation. They include the use of frequency vibrations in Ayurvedic medicine, healing with music, healing with magnetic fields, biofeedback, the laying on of hands, and various forms of mind-body healing.* Acupuncture is another modality—poorly understood in the West—that is used effectively by Chinese practitioners and others. While not perfect, the technique of manipulating subtle energy meridians is more effective than many of the medical and scientific protocols accepted by mainstream Western culture.

---

* The recently created Office of Alternative Medicine within the National Institutes of Health, in a groundbreaking initiative, awarded thirty small grants ($30,000 each) for research in these fields in November 1993. It even sponsored a demonstration of Qi Gong, the ancient Chinese technique of manipulation of the *energeieal* realm, i.e., shattering a rock by the force of mind. Follow-up grants have been awarded to several centers for further work in 1995 and 1996.

Individuals can also channel an unknown healing energy, the most obvious manifestation being the laying-on-of-hands phenomenon. Marcel Vogel, reported in Brian O'Leary's *The Second Coming of Science,* demonstrated that a coil filled with water and hand-charged with this energy can cleanse liquids placed near it of various impurities. In his workshops, O'Leary demonstrates how the passing of hands over and around an individual can enlarge the recipient's energy field, as manifested in the deflection of measuring rods, and increase health and vitality. Psychic surgery (well-documented in the Philippines) is an analog to laser surgery. By focusing and directing subtle energy forces as though they had a scalpel, healers can excise knots of misplaced energy that have resulted in diseased organs.

Medical doctors cite "spontaneous remission" of disease when their healing models cannot encompass the effects of attitude, prayer, or other acts of consciousness. But mainstream research on the use of placebos demonstrates the healing power of belief, as does healing through prayer, which has recently received widespread attention.[15] Research on the healing mind deserves public support instead of censure.

Psychic Phenomena

One of the most promising areas of under-attended research is the field of parapsychology, or psychic phenomena. This area of study portends extensive understanding of the mind-energy-body interaction. New insights here could provide "quantum leaps" in understanding group behavior and fostering healthy social development. For example, U.S. Government intelligence agencies currently fund limited research on remote viewing. Instead of being restricted to defense and intelligence purposes, this effort should be made public and expanded to other institutions.

Another subject for parapsychological research is the phenomenon of cell-to-cell communication, which indicates unexplained transmission of knowledge between species. In addition to an ability to communicate with living plants and animals, humans can exchange information with seemingly "inert" matter. One technique used almost worldwide is that of dowsing to locate water or other subterranean deposits. Dowsing is also used to locate lost persons or objects. Complementing the human ability to receive information in this way is the capacity to *influence* the behavior of material objects. Robert Jahn and Brenda Dunne's mind-impact-on-matter research at Princeton University solidly documents this human ability.

Research on ufology, ETs, and psychic phenomena overlaps to some extent. Some ETs reportedly excel in telepathic communications, hypnotic suggestion, precognition, out-of-body travel, levitation, and psychokinesis.[16] Ingo Swann, renown psychic, has noted that the very fact ETs communicate with humans telepathically proves we too have those capacities.

In each of these areas—ufology, free energy, healing arts, and parapsychology—the biggest unanswered question is the role of conscious intent. The physical laws of matter only account for the propensities or tendencies of things to happen a certain way. What actually makes one of those potentials become manifest is a quantum leap* outside the control of the *known* laws of physics. In these interstices of quantum mechanics, conscious intent or statistical chance collapses the probability wave function; one choice out of many is settled upon by a deliberate or de facto exercise of judgment. Humans can, in specific conscious acts, exercise the power of choice regarding a desired impact and direct subtle energy toward its accomplishment. The scope of our co-creative power to make an "actual event" out of "multiple tendencies" is still unknown. To the extent we have free will, it exists at the level of quantum leaps when we choose one immediate potential direction from among myriad, if not infinite, possibilities.[17]

Revised Framework

Instead of seeking the challenge of leaping into this uncharted territory, most researchers are satisfied to incrementally expand their maps an inch at a time. What is now called for is a new set of theoretical maps with lots of empty space, but with a framework that points to potential relationships between present scientific knowledge and phenomena currently off the chart. In building this framework, *it is useful to extrapolate from our current knowledge*

---

* In physics, "quantum leap" refers to either the movement of particles at a subatomic level from one form to another or the jump across the gap between existence and nonexistence. In human behavior, it connotes acting in a moment of uncertainty without being able to determine the outcome.[18]

base, but not from our current theoretical base. Theories not well-validated, such as the theory of biological origins of consciousness and the deterministic model of behavior, should be suspended to permit a new look at the data we already possess. Concepts and research principles that have proven useful in predicting phenomenal outcomes, whether fitting traditional protocols or not, should be incorporated into a new methodology. For example, Cleve Backster's research on the interaction of human consciousness with plant and animal cells indicates that planned replications (required by current scientific protocols) do not work because the cellular reaction occurs when the experiment is first conceived, not when it is executed.

Taking into account areas of human experience now labeled "paranormal" or "anomalous" requires a new framework that includes the functions of the phenomenal, noumenal, and energeial realms, and the principles of their interaction. The Hermetic Principles can serve as an overarching metaphysical framework to give structure to new experimental maps. As a simple example, the Principle of Correspondence predicts that patterns of force and interaction which exist in one facet of the universe will also inform other facets. This thinking should be applied to the case of the electromagnetic spectrum, reviewing the principle's implications for theory building and research.

Scottish physicist James Clerk Maxwell (who in 1861 summarized equations on the behavior of the electromagnetic force) implicitly used the Principle of Correspondence when he predicted that electromagnetic waves should exist at all frequencies and wavelengths (illustrated in Table VII). Maxwell realized that his four equations pointed to something of a whole, so he tested them in the laboratory. His results showed the speeds of several different categories of waves were all equal to the speed of light. He then hypothesized that if there were waves at "x," "y," and "m" frequencies with the same speed of light, then there must be waves spread across all frequencies (and wavelengths). The result was the discovery of unknown waves. We have not yet identified all the frequencies that may exist on the electromagnetic continuum from zero to infinity, but Maxwell's theory has been confirmed by subsequent scientific, technological, and human experience.

Gaps in knowledge can also be filled by extrapolating from currently known principles. For example, the "passing of hands" through a person's emotional or auric field may be analogous to the passing of an iron magnet over the physical body. With innovative research, we should be able to understand the interplay of magnetic forces with the edge of the *energeia* where emotions affect the body. In mid-1992, the Japanese expanded the conceptual box when they launched a sea craft driven by "magnetohydrodynamic propulsion" (MHP).[19] This technology was based on an extension of current understanding of magnetic fields. Magnetic fields are used to drive a solenoid, generate electricity, and turn motors, where these artificially created fields currently exceed the concentration of force of the Earth's magnetic field. The reported performance of some UFOs (high acceleration, hovering, right-angle turning, and noiselessness) indicates that they may take advantage of magnetic fields. The U.S. Government is working on advanced concepts and engineering to tap this natural force for propulsion systems.*

Ronald A. Brightsen, a former official in the Nuclear Regulatory Commission, has claimed discovery of a heretofore unknown principle in the construction of atoms. The traditional view holds that all atoms are comprised of *individual* neutrons and protons held together by the strong nuclear force. In Brightsen's scenario, powerful infusions of heat—comparable to the dynamics of the sun—are necessary to break the neutrons and protons apart and release the energy for controlled power. (This theory provided the rationale for the multibillion-dollar cyclotron project in Texas that was defunded by the U.S. Congress in 1993.) Brightsen, one of the proponents of cold fusion, contends that atoms are really formed of *groups* of neutrons and protons and that energy can be released by separating these clusters in a more gentle fashion. He claims the flow of electricity through heavy water containing deuterium produces more energy than it consumes. Brightsen and his colleagues in Clustron Science Corporation hope to market this cold-fusion technology for the generation of electric power.

Since all sections of the electromagnetic spectrum move at a constant speed (186,282 miles per second)—the "c" in Einstein's formula $E=mc^2$

---

* According to the British press, John Searl of England demonstrated such a propulsion system in a "flying disc" 25 years ago, but the technology was allegedly suppressed by the British government. Like the United States, the United Kingdom has a history of significant covert government research and corresponding cover-ups.

—the Principle of Correspondence would predict that other spectra exist at other constant speeds (see Table IX). At this point we can only speculate as to the inherent ratio, but undoubtedly another parallel spectrum (or more)—running at a different factor of electromagnetic speed—exists. It affects our universe, and therefore our lives. Nikola Tesla* claimed to have detected another spectrum of waves—he called them "cosmic waves"—moving at fifty times the speed of light. Recent experiments by Tom Bearden (see Chapter 2) are purported to have resulted in wave velocities of up to eight times the speed of light.

TABLE IX

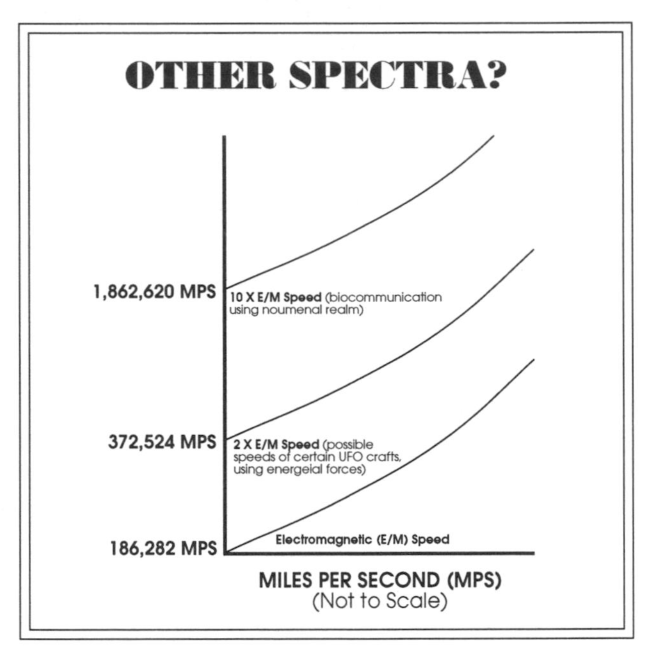

What tangible forces, susceptible to testing, could be involved in such yet-to-be-identified spectra? The first candidate could be the force involved in cell-to-cell communication—the hypothesis being that, over distance, living

---

* Nikola Tesla, the inventor of alternating current and wireless radio almost a century ago, developed a device now known as the Tesla Coil. This coil functions at the interface of form, energy, and thought (read *intention*). Derivatives of this pioneer's work include free energy motors, gravity field technology, and zero-point energy.

cells exchange certain kinds of information faster than the speed of light. Although Cleve Backster's experiments (described earlier) seem to indicate instantaneous communication, the distances involved in his experiments are too short to allow accurate measurement of such speeds. In the three-aspect model presented here, the only theoretical limit to the speed of information transmission is the capacity for cognition.

Research involving cell-to-cell or biocommunication would require distances sufficient to measure speeds of perhaps more than one million miles per second. The hypothesis could be tested by sending human white cells on interplanetary vehicles and correlating, by use of time-synchronized instruments, the actions of the donor with respective cell reactions. This experiment would provide the necessary units for calculating the speed of information transfer. The findings could have implications for understanding subtle phenomena such as telepathy, clairvoyance, or remote viewing. Similarly constructed experiments, like those involving the splitting of photon pairs, could add to our knowledge of such information exchange. The "Spindrift Experiments," conducted by Christian Science researchers, examined the mind's influence on seed germination, yeast, and healing.[20] Their research is deserving of replication.

The Hermetic Principles may offer theoretical insight here. Such a gentle force as Brightsen's is likely to exist only in a polarity where a strong force and a weak force operate in reciprocity. The subtle force that animates emotion is likely to have a spectrum of vibrations with lower frequencies of hate and anger and higher ones of altruism and joy. Will one end of the gradient manifest cellular damage, energy congestion, and ideational conflict, while the other provides healing, gentle flows of feeling, and harmony? How these unidentified forces and biocommunication relate is another gap in our knowledge. We must reconceptualize the known and integrate it into a larger hypothetical framework with gaps for testing, questioning what we now consider absolutes and seeing them in new relationships. The following section illustrates one such path, the development of a useful *metascience*.

Defining Metascience

Given their limited nature, the four forces—gravity, electromagnetism, weak force, and strong force—must exist in the context of a larger framework. Scientists have deduced these forces from our experience in and analysis of the phenomenal realm. We know apples fall from trees; therefore, gravity exists. We know iron filings gather around a charged iron bar; therefore, magnetism exists. We know stored current turns motors; therefore, electricity exists. When we sit in a chair without falling through it, we can believe some strong force is holding it together. When we hear a Geiger counter click and see our teeth in X-ray film, we can infer that some weak force permits atoms to separate. (See the Chapter 2 description of the subatomic nature of these forces.)

Scientists tell us that *everything* that happens in the universe is due to the functioning of one or more of these forces. (Recall the Grand Unifying Theory search described in Chapter 1.) Such a conclusion is premature. These forces cannot account for the continuing process of reconstruction (creating and recreating) that counterbalances the phenomenon of entropy and characterizes the evolving universe. They offer no insight into the larger questions: What force caused the birth or rebirth of the universe? Why is there directionality within the system? Are the patterns that shape the ebb and flow of matter and conscious life inherent in creation? Are they subject to modification within the universe? Is the arrow of time really a spiral? Why do the four forces not explain many human experiences (as highlighted in Table X)? The reality seems to be that these four forces may be the least tractable in the overall schema of the universe. They are much less susceptible to direct manipulation by the mind than those of the subtle realms. We need a *metascience* that incorporates all conceivable forces and their interactions.

What we know about the four forces is limited by our current scientific tools. Each force impacts the material realm through its own particulate matter. Gauge particles are exchanged between two objects, giving rise to mutual attraction, like heavy balls being hurled from person to person. Of course, the gauge particles are only mediators of the energy, *not the energy itself*. The elusive energy in turn is patterned by some designing force. Could it be thought? Gravity is assumed to work through a hypothetical particle called the graviton, although *scientists have not yet found one*. Puthoff, et al., argue that gravity is simply another aspect of the electromagnetic field.[21] Photons, having neither mass nor charge, travel at the speed of light and seem to be the mediators for the electromagnetic force. The strong force is mediated by eight types of gluons (appropriately named) and the weak force by two particles—a "W" and a "Z."

A *metascience* positing other forces and phenomena illustrates a theoretical approach that is susceptible to empirical research. Its extrapolations are an attempt to relate different levels and categories of unexplained human experience to the current scientific baseline. Table X suggests two directions for new hypotheses that expand our analytical vision. One is that the full range of influence of the four labeled forces is not yet understood; an undefined part of the electromagnetic spectrum could account for some of the events now labeled paranormal. For example, magnetic field effects

Table X

ONLY FOUR FORCES?

Types of Forces

Phenomena	Gravity	Elec/Mg	Strong	Weak	Unknown	Unknown
Light		X				?
Sound		X			?	
Movement	X					?
Matter			X	X		
Magnetism		X	X		?	
Materialization			?			?
Telesenses		?			?	
Mind-Brain				?		?
Healing Hands		?			?	
Emotional Bond	?					?
Bio-communication		?			?	

A search for GUT is premature

are much more extensive than earlier imagined. On a mundane level we are just learning of the impact of changes in the magnetic field (caused by electric razors and hair dryers) on cancerous cell growth. Slowly, recognition is being given to the relationship between cancer rates and high-voltage transmission lines. Do the magnetic field and the mind-body organism interact in a way that changes the functioning of the cells, causing them to react differently to external stimuli, including carcinogens?

A second research question for *metascience* is whether forces exist in the realms of *energeia* and *noumena* that are parallel to the currently identified four phenomenal forces. Given Hermes' insights, the forces are likely to relate to the dynamics of attraction/repulsion, bonding/disintegration, two-way communication, and realm-to-realm transformation. Table X is only a starting point. Unexplained experiences, including such mysteries as events that validate astrology and prayer, have to be fitted into this schema. Physics is now in a cul-de-sac of incoherent over-elaboration, with a literal maze of particles and forces. A larger *metascience* could help to bring about needed order.

The placing of the four forces within the framework of a Hermetic-based *metascience* offers some intriguing areas for mind games and laboratory research. The Principle of Polarity can be applied to the forces of gravity and antigravity: We need theories that incorporate explanations of antigravity and antimatter in relation to their polar opposites. The strong and weak nuclear forces appear to represent polarities of the same bonding principle, but what is the polarity of the electromagnetic force? Or is the magnetic force the polar partner of the electrical force? If so, where are the magnetic monopoles? (In anticipation some scientists are actively looking for them.) We need to study the flickering dynamic that functions between energy and mass, between thought and subtle energy, and between general consciousness and group behavior. Only then can we begin to understand the synthesizing process that gives rise to organic life from the void.

All the physical forces—through the infusion of patterns that collapse them into vectors and form—seem to be susceptible to the influence of consciousness. They, like their obverse particles of matter, are imbued with vibrations subject to measurement. The forces themselves, unseen by us, exhibit the masculine expressive gender; the mediating particles are the feminine receivers that embody and sustain the forces. Other Hermetic Principles are exemplified in the following two examples.

Eric Dollard, following up on some of Tesla's work, has identified naturally occurring patterns (termed the Golden Ratio Spiral) in various kinds of energy release: a waterfall, a crack in glass, and water percolating through sand. The patterns seem to fit the wavelength and frequency of the energy discharged, bursting forth and then dissipating. Given this data, he has inferred a more general principle at work. Is this a function of the Hermetic Principle of Rhythm? Is that same principle relevant to the patterns of personal growth and social change? If so, it is possible to become more conscious and in control of them.

Russian scientists, not bound by the same conventions as their American counterparts, extrapolate more quickly from established thinking. Professor Alexis Zolotov,[22] for example, postulates three palpable energy fields parallel to the biological one that is generally accepted in the West: electromagnetic, cinetic motion, and information. Each of these, Zolotov believes, has carriers and vectors that affect all bodies, whether a person or a planet. His ideas of the biological, energetic, and auric/information fields correspond roughly to the *phenomena*, *energeia*, and *noumena* posited throughout this book.

Although institutional science has not yet validated the existence of something like an information field, psychics and healers access it and use it. Rupert Sheldrake popularized the concept of a "morphogenetic field" that can be developed over time by conscious intent and practice until it achieves a living force of its own.[23] The field's pattern of force can then affect the behavior of other beings as they come to resonate with its frequency.* Based on this concept of interfield communication, researchers can devise experiments to test Zolotov's and my ideas of how these fields interact with matter.

As the scope of a *metascience* broadens, we will find answers to many current questions that now require blind (usually misguided) faith for answers. We will develop a theoretical base for the obvious links between

---

* The popular, though possibly apocryphal, "Hundreth Monkey" story illustrates an amazing phenomenon: the learning in one group of monkeys is somehow absorbed by a separate, distant group of monkeys.

thought and the production of matter in the nervous system (Principle of Mentalism). We will identify a scale for measuring the bonding of individual and group thoughts to subtle energy, in turn creating emotions that affect matter—as in the physically empowering effect of distant prayer and the physically deleterious effect of existential fear.

While John Horgan, in his book, *The End of Science,*[24] concludes that science has reached its limits, he accepts the current major scientific theories as a "modern creation myth" that will hold true for a thousand years. That myth (in this case, unproved belief) includes the Big Bang, the four forces, the chance evolution of life, and the biological determinism of behavior—all of which stand as clearly partial explanations of experienced reality. A metaphysical perspective such as the Hermetic Principles points the way to a more comprehensive approach to scientific research, beyond the current levels of biology and physics. Such *metascience* has practical implications for a much broader range of human understanding and development.

NOTES

1. Zukav, Gary. *The Seat of the Soul* (Simon & Schuster: New York, 1990).
2. Bateson, Gregory. *Mind and Nature: A Necessary Unity* (Dutton: New York, 1979); Talbot, Michael. *The Holographic Universe*.
3. Capra, Fritjof. *The Tao of Physics*.
4. Jou, Tsung Hwa. *The Tao of Tai-Chi Chuan* (Tai Chi Foundation: Warwick, New York, 1981).
5. Ackerman, Diane. *A Natural History of the Senses* (Vintage Books: New York, 1991).
6. Several species have been shown to have a definite orientation to magnetic fields, but the impact of magnetic forces in human information processing is not yet clear. In *Human Navigation and the Sixth Sense* (1981), Robin Baker of the University of Manchester in England reported less directional certainty among blind-folded volunteers with magnets on their heads than those blind-folded and without external magnets.
7. Leadbeater, C.W. *The Chakras* (Theosophical Publishing House: Adyar, India, 1927).
8. Hopes, David. *Sun*. May 1992.
9. Mack, John E. *Abduction: Human Encounters with Aliens* (Charles Scribner: New York, 1994).

10. Haines, Richard. "Fifty-Six Aircraft Pilot Sightings Involving Electromagnetic Effects." *Proceedings*. (MUFON Symposium: Albuquerque, New Mexico, July 1992).
11. Howe, Linda M. *Alien Harvest.*
12. O'Leary, Brian. *Miracle in the Void* (Kamapua'a Press: Maui, Hawaii, 1995).
13. *"Cold Fusion" Magazine*. Peterborough, New Hampshire.
14. *Proceedings.* (Institute for New Energy Conference: Denver, Colorado: May 1994).
15. Dossey, Larry. *Healing Words* (Harper & Row: San Francisco, 1993).
16. Fowler, Raymond E. *The Watchers* (Bantam Books: New York, 1991).
17. Wolf, Fred Alan. *Taking the Quantum Leap* (Harper & Row: San Francisco, 1981).
18. Heisenberg, Werner. *Physics and Philosophy* (Harper & Row: New York, 1958).
19. MHP takes advantage of the Lorentz Force (named after its discoverer, Dutch physicist Hendrik Lorentz, a winner of the 1902 Nobel Prize). Used in motors and pumps, the Lorentz Force essentially involves harnessing the perpendicular force that results from two vectors of an electric current and a magnetic field positioned at right angles to each other. The movement of the electrically charged particles pushing against the magnetic field results in an opposing force that propels the sea water through a pipe. The result is thrust like that driving a jet plane. A similar technology, using gravity as a countervailing field, could be developed to achieve greater speed for airships than is possible with air as the medium for jet propulsion.
20. Owen, Robert. *Qualitative Research: The Early Years* (Grayhaven Books: Salem, Oregon, 1988).
21. Haisch, Rueda, and Puthoff. "Beyond $E=mc^2$."
22. Vallee, Jacques. *UFO Chronicles of the Soviet Union* (Ballantine Books: New York, 1992).
23. Sheldrake, Rupert. *The Presence of the Past.*
24. Horgan, John. *The End of Science* (Helix Books/Addison-Wesley: Reading, Massachusetts, 1996).

Chapter 5

Process: Causative Mind

For millennia religions have assumed other realms existed outside of and beyond the reach of daily life. Modern science doesn't believe in them at all, preferring to label as "anomalous" the human experiences that lead to such beliefs. But how do we account for dreams and other alternate states of consciousness? What about evidence of communications from the dead and other beings outside our physical space? How do we explain the influences of thoughts on material objects? Where do memories reside? Quantum physics and consciousness research offer explanations and help reconcile the two belief systems. Sentient beings serve as the synthesizers of all realms.

WHILE SCIENTIFIC THINKING HOLDS that all action of importance takes place in the material realm, human experience proves otherwise. Analysis of each phase in cosmic life (birth, struggle, joy, creation, growth, attraction, pain, death, etc.) reveals a functioning multilevel reality. For beings to perform at their potential, they must be aware of and open to the interaction of the *phenomena*, *energeia*, and *noumena*. Ignoring any part dims the fullness of life. Since the beginning of the current historical era on Earth, less than 5,000 years ago, humans, while elaborating various technologies, have regressively reduced their scope of knowledge.

The spotlights of Western science and religion draw attention to certain fragments of experience, while ignoring important others. Eastern philosophy is just as remiss. Many, whether Hindu or Muslim, share a perspective that is as dualistic as the Cartesian view that differentiates between the physical universe and the hidden plane: they claim the former is knowable and discoverable by science, but the latter is separate and accessible only through, respectively, Yogic meditation or the inspired Koran. But there are no such impermeable barriers. The evidence reviewed in this book indicates that the subtle realms usually thought of as beyond humanity's reach are not really "beyond" after all. Heaven and hell, spiritual eternity, the logos—are all omnipresent throughout the cosmos. Regardless of the nature of the experiences—extraterrestrial, extradimensional, outside history, or beyond the waking state—they are all aspects of the seamless, conscious universe.

The discoveries of new science and parapsychological research, including the antics of a Tomaz Green-Morton in Brazil,[1] who performs materializations while drinking rum and blaspheming, demonstrate the singular and neutral nature of universal principles. Neither adherence to a specific scientific protocol nor a particular religious orientation is necessary for access to another realm. The best route to understanding the true nature of our circular universe is *metascience*, a combination of all ways of knowing that can be eventually tested in the phenomenal realm. *Metascience* places access to knowledge in the hands not only of specialists, but also of independent individuals willing to test their hypotheses in all realms. The following discussion illustrates how consciousness permeates the realms of subtle energy and *matenergy*, and serves as the unifying force for every aspect of universal life.

Memory: Cause and Effect

Even though the *noumena*, from a conceptual perspective, is distinct from phenomenal reality, in existence and function it is not a realm beyond. As we have seen, the *noumena* and *phenomena* are synthesized through the *energeia*, the bonding and transforming force. As integrated parts of the whole, all three have mutually dependent functions, affecting the way beings react to mundane affairs. The nature of their ongoing interaction gives continuity to life forms and their behaviors. Such continuity is sometimes labeled *memory*—cellular, muscular, energetic, mental, etc.

Memory results from the interaction of the three facets, *and* serves as the glue to hold them together. Images and ideas live on in the *noumena* (as in memory) in recognizable form, as long as conscious beings charge them with sufficient emotional energy, from the *energeia,* to keep them intact. Passing thoughts or comments, unsupported by additional emotional investment, dissipate over time. Then even the most finely tuned brain cannot retrieve them for a searching human mind.

Unless words and concepts generated by a conscious being are invested with adequate emotional energy, they quickly fade from memory into the noumenal background. (When there is a subtle energy boost, it appears to be reflected in strengthened neurotransmitter connections.) The primal *logos* or schema of the universe could have easily evaporated, had the Grand Couple not infused it with sufficient intentional energy to impel continuing manifestation in the phenomenal realm. A passing fantasy on a lazy summer afternoon never finds its form in the subsequent course of events, while a creative flash, powered by high motivation, can result in a new mechanical invention or social structure. Strong ideas combined with emotion can even affect the performance of machinery, at the time of its creation or at a later date.

When a being or group charges an idea with a strong emotional response, that idea becomes a personal and noumenal memory, remaining available to open and receptive minds throughout the cosmos. Researchers who know nothing of each other's current work frequently discover later they were working on the same concept at the same time. In the late 1980s, the idea of democracy remained vital, while that of communism quickly waned. Communism's thought forms in the *energeia* had a half-life (rate of decay) that was in inverse order to that of human freedom and self-determinism. This relationship of the emotional charge to decay rate is why the great ideas of all time still resonate today and appear in dreams, meditation, or creative free-thinking. They are not just stored in material data banks, but continue to have a long life in the *noumena*—parts of which were termed the "collective unconscious" by Carl Jung and the "Akashic records" by Rudolph Steiner and others.[2] ("Akashic" is derived from the Sanskrit word *Akasha*, meaning all-pervasive space.) Enduring memories, contributed by all conscious beings, from the Grand Couple forward, literally hold the universe together.

The decision to charge an idea with sufficient energy to make it survive can be either a deliberate or quasi-autonomous response. When thousands of people *deliberately* spend one hour a day meditating for world peace, entire societies can feel the influence in the noumenal realm.* On the other hand, when millions respond *spontaneously*—but deeply—to the beauty of an image such as Earthrise as seen from the moon, the thought *(noumena)* form is charged with a force *(energeia)* that will power decades of emotion and action. Deliberate charging explains how conscious imagery can be used over time to affect one's health and interactions with others, while the impact of one dramatic emotional experience demonstrates how a religious healing can work.

Given the Principle of Correspondence, in an integral universe, the dynamics of the cosmic *noumena* can be inferred from the workings of an individual's memory. In this context, the longevity and impact of an idea or symbol in either individual or noumenal memory would depend on the clarity of the symbol, the level of emotion bonded to it, and its subsequent recharging by one or more beings. An idea generated by a computer and never read by a person would not interact with the *energeia* and the *noumena*. Only symbols that have been charged, either positively or negatively, by the emotions of conscious beings can have a direct effect on the phenomenal world. The Kahunas of Hawaii understand these principles; they teach that a prayer will receive a response only when it is offered to the higher realm *(noumena)* with all the personal psychic force one can muster.

How the brain assists in the acquisition, storage, and recall of specific images (whether words or pictures) is not clear; however, scientists have been able to observe the transmission of electrical and chemical impulses by neurons. Neuronal reactions to experience (internal or external) seem to be correlated with the quality and level of emotional excitation. Personally experienced events affect more areas of the brain than do the second-hand acquisitions of memories. While the neurons, electrons, and neurotransmitters

---

* Some believe that the efforts of Western meditation groups contributed to the force of President Gorbachev's vision of *perestroika* in the former Soviet Union.

that make up a physical event are transitory, specific memories seem to have their own relatively permanent existence. They can be transformed or modified, but do not appear to disappear quickly. How long memories can last in the *noumena* is not known, but some memories from prehistoric periods can still be accessed through dreams or deliberate remote viewing.

Humans can consciously access memories as the events occur or after the fact. In a fateful example of the first category, the U.S. military had access to subtle-sense knowledge of Japanese ship positions as they moved toward Pearl Harbor (and later in the war as well). A lieutenant-commander who was a code specialist in the U.S. Navy gave this and similar information to his superiors, but it was discounted because he gained his information through "intuition," as opposed to deciphered radio codes. He knew how to decode images of current events in the *noumena*. Of a more permanent nature are the previously mentioned Akashic records. Analogous to a video recording of an event in the five-sense realm, the Akashic record retains the noumenal history, serving as the memory bank of human experiences, including interactions with other beings. The mental creations collectively making up the *noumena* may be as substantive and permanent as anything in the universe, yet they are always available to the subtle senses.

Some of the more permanent, culturally-based memories influencing behavior were labeled "archetypes" (Greek for original pattern) by Carl Jung. These patterns in the *noumena,* reinforced by conscious beings over generations, are bonded with enough subtle energy that they continue to exert subliminal influence over the ages. Archetypes like "hero," "victim," "journey," and "hunter," indirectly help shape individual lives until they are brought into awareness.

Large-scale morphogenetic fields (described in Chapter 4) are another illustration of how collective memories from the realms of *noumena* and the *energeia* shape matter and behavior.[3] In this process a specific idea becomes so energized by a critical mass of people holding it that it affects the thinking and behavior of countless others. Memory here plays the role of cause and not effect. An illustration: The impact of Cold War ideas on emotions and behavior, as far removed from the facts as the stereotypes were, can best be explained in terms of the *noumena* and *energeia* where they manifest as energized thought forms. (In a smaller scale indication of their power, individuals whose homes have been the site of intense emotional battles have reported the ongoing influence of negative thought forms.)

It is impossible to argue that telecommunications, since they did not exist, were responsible for the similar nature of the American and French revolutions of the eighteenth century. But the creations of a few writers, the noumenal constructs of liberty, equality, and brotherhood, being fully charged in the *energeia* by masses of people on one continent, could motivate people energetically on the other side of the world. That energetic connection (not unlike the hundredth monkey theory mentioned earlier) could help account for the fervor of citizen armies.

Perhaps the patterns of political upheaval that traversed the globe in 1968 and 1989 were less influenced by television than the power of noumenal fields containing highly energized patterns of political thought. It is also probably true that telecommunications media and computer networks, by giving more people access to the same images and ideas, help the thought forms accrue more subtle energy, extending and strengthening their force in the *noumena*. Independent thinkers wake up each day to "discover" they share new experiences of physical and psycho-spiritual reality.

Millennia ago, Chinese scholars recognized how fields of tendencies, or incipient noumenal patterns, impacted the course of individual and collective events. They observed how the confluence of thoughts and subtle energy forces affected the lives of individuals. Knowing it was possible for a person to divine the force fields relevant to personal choices, they developed a process for obtaining this intelligence, The *I Ching*.[4]* The principles underlying the *I Ching* were based on an appreciation of the interconnectedness of ideas, energy, and conscious action. Ancient Chinese culture understood a consciously lived life is an orchestrated nexus of all realms. Living with the *I Ching*, the Runes, and other ways of tapping inner knowledge is like being in the flow of Carl Jung's synchronicity where the desires and needs of the many are mediated through the invisible medium of the *noumena*.

* It provides for the selection of relevant advice on an issue from a book of wisdom where the selection is controlled by subconscious divining of symbols.

Other Sources

The interaction of the *noumena* with daily life involves more than human ideas and communication. The communications of all beings involve the *noumena*. There is growing recognition that all animals react to humans, and that the condition of plants and the earth itself gives them feedback on their behavior. Many now know the interaction can be two-way and positive as well as negative. Some are coming to realize that more acknowledgment of the interspecies exchange of information and subtle energy could enhance the quality of all lives.

People can learn when they suspend intellectual prejudgments, as in the following example. In the fall of 1988 in Barrow, Alaska, three gray whales became trapped beneath an ice hole five miles from open sea. Although a series of new holes had been created by scientists and local Inapt people, all efforts to entice the whales toward the new holes and the sea beyond failed. Finally, when Jim Nollman of Interspecies Communications arranged to send music through the hole the whales were currently using (which was about to freeze over) and also through one of the new holes, the whales moved to the new hole. The whales immediately started to migrate down the channel, surfacing to breathe at new holes along the way as music was played into each one.

The whales understood and acted in relation to human-generated music in a way that astonished onlookers who had no explanatory frame of reference. The indivisible *noumena* and *energeia* help explain the obvious multichannel nature of such intra- and interspecies communications.

Interspecies communication exists not only among humans, animals, and plants, but among other conscious beings throughout the universe. Many individuals who encounter another form of being automatically attribute it with godlike wisdom and/or demonic powers. The contactee frequently believes he or she has had a special encounter with a divine source. But the early Egyptians, Greeks, and Romans had a clearer understanding: they saw the beings with all their foibles, as well as their insights or powers, as cohorts in the same field of consciousness. Their myths imply many other categories of beings were involved in similar life cycles and conscious development.

Just as perception of the electromagnetic spectrum is partial, knowledge of other realms is very limited. Being wedded to a mundane concept of space-time can interfere with understanding other aspects of the *noumena* and *energeia*. For example, the noumenal realm appears to involve an interspecies flow of information transcending the speed of light. (Recall the experiments with cell-to-cell communication presented in the last chapter.) Subtle energy forms, while sometimes tied to a physical form (as in a person's auric field measured by Kirlian photography), appear not always to be constrained by space-time (as in out-of-body experiences or OBEs).

The idea that noumenal and energeial forces exist outside the currently defined space-time universe is not farfetched. Newton explained there is no absolute space (everything is in constant movement), and Einstein's theory of relativity ended the notion that ordinary time is an absolute standard, uniform throughout the universe. These understandings make plausible the existence of reality outside the field of our typical five-sense perceptions. In the 1960s Stephen Hawking and Roger Penrose proved mathematically that *time as we know it* had a beginning. If one accepts that proof, it becomes equally credible that other realms could have different beginning points. It is just as reasonable that an overarching cosmic consciousness should then be able to bridge the membranes that separate these parallel realms.

If people are to assimilate information from a wider spectrum of reality, they must reject not only a finite conception of space-time, but also the narrow view of consciousness promulgated by current science. As Robert J. Hannon aptly states, "science" can be seen as the domain of a group of professionals who sometimes decide what is "true" and what isn't before a hypothesis can be proved.[5] As an example, he challenges the assertion that the red shift (the earlier mentioned fact that objects moving away from us appear to shift to the slower frequencies of the color spectrum) proves the continual expansion of the universe, calling it unsubstantiated. Similarly, when priests assert unproven dogmas that demean alternative interpretations, they too limit the search for wisdom. A more open-ended and adventurous, yet rigorous, approach to hypothesis development and research is needed.

Unified Consciousness

Experiments with mind travel, such as OBEs, remote viewing, and telepathy, appear to confirm the existence of a unified field of consciousness and lend credence to the view that humans are integral partners, not serendipitous bit players, in the cosmic drama set into motion by the Grand Couple.

Spiritual texts and legends not based on a singular reality assert the existence of various realms inhabited by substantially different beings. Simplistic Christian cosmology has three such realms: Heaven, Earth, and Hell. Catholicism posits two in-between realms: Purgatory, where the soul is purified before it is ultimately admitted to Heaven, and Limbo, where pure souls await assistance to ascend. In very general terms, the Nirvana of Buddhism is comparable to the Christian Heaven, while Bardo parallels both Purgatory and Limbo as stages in the cycle of reincarnation. An afterlife, in the religion of Egypt, was believed to take place beyond the river of the dead. Individual access to all such kingdoms was believed to have been controlled from above. These and other religions were founded on the belief that a god from that higher realm either 1) bestowed divine status upon a chosen Earthling, or 2) incarnated itself in human form to provide an exclusive channel between the different realms. These divine men and women allegedly produced miracles and shared visions that followers took as evidence of their divinity.

Now there is reason to question such claims of separate divine realms. The presumed divine beings may have been ordinary conscious entities whose miracle-producing skills were simply technologies beyond the human observers' experience. Some of the most comprehensive and best-researched evidence of this is from the previously mentioned *Earth Chronicles* (four volumes) and *Genesis Revisited* by Zecharia Sitchin.[6] His profound synthesis of many disciplines, with well-documented evidence involving numerous scholars over two centuries, supports the thesis of Earth's occupation by beings from elsewhere in the cosmos. Sitchin's analysis indicates one group, the Annunaki, was from within this galaxy, with comparable mental powers, once again pointing to a singular pool of consciousness that includes humans.

Perhaps even more compelling evidence of a singular consciousness are the *current* activities of similar conscious beings on Earth (and possibly elsewhere in the solar system). In May 1992, the Rocky Mountain Research Institute and the International Association for New Science convened a forum in Colorado for the most comprehensive review ever of the evidence for extraterrestrial intelligence. Presenters at the conference demonstrated to the satisfaction of many skeptics gathered there that not only are other types of beings active in different parts of the universe, but that some of them can evidently transcend the familiar dimensions of our space-time. Evidence included photographs, artifacts, traces, human/ET contact case studies, and personal experience.[7]

A wider conception of the unicosm is called for if we are to encompass the so-called "alien experiences" of a cross-section of humanity. In addition to humanoid ETs, who are the likely basis for the "gods of old" legends, many humans have encountered beings in disembodied form. Known as ascended masters, angels, spirit guides, or guardians, they speak to people directly, or are channeled through an individual in a trance who serves as their voice. There are reports of energeial beings channeling messages through computers and telephone answering machines, or other voice recorders. Seen as a source of wisdom and general insight, such "masters" also offer specific communications on departed relatives or stock market trends. The emergence of extensive channeling and the expression of "miraculous" powers (such as materialization, psychic, and healing powers) among the noninitiated has undermined the notion that communication with higher levels is restricted to a chosen few.[8]

Also providing additional evidence of human access to a unified consciousness are individuals who deliberately project their minds and energies into the nonphysical aspects of the universe. Instead of sending specific thoughts into the *noumena*, the whole mind goes. Consciousness travel requires knowledge of the process and some practice, like that employed by the systematic researchers of the Monroe Institute in Virginia.[9] The late Robert Monroe assumed that individual consciousness is not bound by its corporeal home and discovered a gateway through which an individual consciousness can deliberately move between realms. The descriptions of Monroe's consciousness travelers indicate they experience the interstices of the *energeia* and the *noumena,* the linkage of subtle senses where consciousness is aware of different energy forms.

Participants at the Monroe Institute are placed in a Holistic Environmental Chamber that isolates them from light and some degree of electromagnetic radiation. The simultaneous feeding of delta sound waves to the cerebellum and theta waves to the cerebral cortex (called Hemi-Sync, for *hemispheric synchronization*) induces an OBE for the participant. The conscious travelers—in a state of "body asleep/mind awake"—report meetings with other entities, views of distant places, and awareness of a vast realm of consciousness other than Self. However, the traveling Self always sees itself as a part of that whole.

Acknowledgment and discussion of the above-described variety of communications would lead, one hopes, to serious mainstream study and scientific exploration. The following sections illustrate potentially fruitful areas for research, involving a number of ways in which the universe's unified field of consciousness is expressed: the role and function of dreams; the event we call death; parapsychology in the unbroken gradient of consciousness; communication with other beings; and interdimensional shifts within a spectrum or among distinct spectra.

Dreaming

Alexander Borbely, in a comprehensive review of modern scientific research on sleep, concluded: "Not only are we investigating a process that usually occurs in the dark, but we are also almost completely in the dark about its function."[10] During sleep, dreaming—a natural and obviously important human function—is one of our most important channels to cosmic consciousness.

The physiological effects of sleep can be described. It can be artificially induced. Humans can manipulate its patterns, measure its electrical and muscular impact, label its different levels, but they cannot prevent it indefinitely. Scientists have not detected all its specific benefits, nor do they understand why and from where the dream images come during sleep.

Looked at from the perspective of undifferentiated consciousness, these images seem to be caused by stimuli from the general field of consciousness: coming largely unfiltered during physical sleep, the input leaves the individual free to make interpretations. In dreaming, an individual's vibrational state sends out certain signals and, like a radar, subtle senses receive the returning vibrations as squiggles on the internal screen. The blips give the direction, force, speed, and magnitude of the signals, but do not directly reveal the nature of their source. The patterns require a mental act to interpret the data and decide on the response. The two differences from the waking state are that in dreaming the subtle senses take precedence over the physical senses and the degree of censorship is reduced.

However, even though the personal membrane of consciousness is permeable, it always remains subject to at least the partial control of the aware being. Dream material, like all other sensory information, can only pass through with explicit or de facto permission. People are now learning to monitor and intervene in their own dreams. Each being at some level is its own dream keeper, even when it has committed acts—such as the ingestion of chemical substances—that render it incapable of rational control.

Like the gradations in the waking state, dreams in sleep are a composite of various degrees of awareness. Several hours observations of anyone's waking state reveal periods of concentrated mental focus, followed by moments of distraction when unbidden thoughts creep in. There is every reason to believe that similar variations occur in the sleep state.

When one is awake but in a relaxed state, having suspended conscious effort at problem solving, breakthroughs frequently occur. Paradoxically, *more* overall attention is focused on the problem than is possible in the midst of distractions created when all senses are highly active. This dynamic is why dreaming, like controlled entry into meditative states where much sensory input is shut down, is so helpful for creative effort. It is often used by people who feel blocked; they articulate their questions and leave the mind free to search for the answers during sleep. Material can be perceived during such "conscious dreaming" from two almost unlimited sources, both of which are also partially accessible in the state of wakefulness: telepathic access—clairvoyant reception of current information; and Akashic access—tapping into the *noumena* of past information.

Beyond Death

Near death experiences (NDEs) and death itself provide further evidence of an undifferentiated consciousness. As mentioned earlier, almost every culture believes in a realm where the spirits of recently deceased persons dwell, at least temporarily. Such beliefs are rooted in two planet-wide human experiences: the common perception of a continued nonphysical presence of those recently deceased and specific and verifiable communications between individuals and their "dead" relatives.[11] Thousands of people, having been resuscitated after "clinical" death (the cessation of respiratory and brain functions), report similar attributes[12] of a nonphysical realm inhabited by those who have died.

Attributes of the non-physical realm
• Ability to hear conversations in the ordinary realm
• Feeling of peace and calm
• Various kinds of noise (from harsh to musical)
• Passage through a tunnel (usually dark)
• A sense of a nonmaterial body
• Meeting familiar others
• Encounter with beings of light
• Review of life's significant events
• A transition point (door, fence, fine mist)

NDEs and communications between the dead and living clearly demonstrate that personal consciousness ranges across realms and simultaneously perceives the *phenomena* and the *energeia*. They offer validation for the three-faceted model of reality posited in Chapter 4.

In the context of the assumptions in this book, the term NDE, as popularized by persons such as Raymond Moody and Kenneth Ring,[13] appears to be a misnomer, because *death* may be the wrong term. The realm of the NDE experience, which involves beings we call dead, is not unlike that experienced in the "travels" of people at the Monroe Institute. Both involve a shifting of conscious awareness to the realm of the subtle senses, which is accessible to all beings. Finding death in life and life in death reveals that both are living experiences.

The experience of "death" itself, then, may be only a particular form of OBE. During both OBEs and NDEs, one senses one's consciousness leaving the body and moving to another locale, yet remains aware of what is going on in both realms. In contrast, a true death would be a return to the complete dormancy of the preorganic void, a state beyond current comprehension and apparently one humans manage to escape through the universe's cycle of rebirth. Plato may have had it right when he described birth as going to sleep (and forgetting) and death as awakening (remembering). Death, or de-incarnation, with greater access to the *energeia* and *noumena* and free of phenomenal boundaries, gives an entity greater liberty to roam the universe.

The *Book of Revelation*, *The Tibetan Book of the Dead*, and *The Egyptian Book of the Dead,* when read in the light of this new perspective, offer insight into the permanent departure of consciousness from the physical body. In describing a phased process of disengagement, they refer to a period of life review and judgment. Free of the body's constraints on memory, the being can more easily recall the experience of many lifetimes, and assess how well it has lived this one. This possibility is compatible with the traditional concept of reincarnation. The study of NDEs and OBEs and their relationship to other so-called paranormal, or energeial, experiences may reveal how a being can consciously shift among different states during a lifetime.

Unbroken Gradient

The unbroken nature of the circle of cosmic consciousness is also demonstrated by the lack of a definitive line between an OBE and so-called parapsychological experiences. What starts as a remote viewing experience, initiated in a deliberate manner, may end up as an involuntary OBE. Where the "paranormal" experience remains on the spectrum of conscious states appears to be largely a function of intent. For example, the difference between remote viewing and flashes of telepathy is only a matter of degree. The former is a conscious act and the latter is more spontaneous, but they operate on the same basic principle, drawing on the collection of current images in the *noumena*. Precognition is also an exercise of the same subtle sense, except that it reaches forward into space-time and perceives the potential for events to come. Hypnotic regression reverses the same process, revisiting the past on the arrow of time.

All these communications are different ways to access aspects of the *noumena.* Each tunes the frequency of our subtle senses to an appropriate energetic range, somewhat analogous to physically accessing a radio frequency of the electromagnetic wave spectrum.

Despite popular time-travel fiction and reports of government research projects like the Philadelphia Experiment, from this space-time, it appears an incarnated consciousness can neither go back and change past events, nor directly manipulate the future from the present. The functioning of cause and effect means that a change in the past would change the present. Therefore it is impossible to go from the status quo of the moment and change the past without modifying the present in a perceptible way. The resulting oscillating loop would become perpetual, breaking down the directionality of space-time.

In *The Watchers,*[14] a nonfiction report of an abduction, alien beings remark, "We can know what is going to happen but we cannot intervene." That appears to be a statement of universal fact, at least as pertains to humans and similar conscious beings. An account in the *Washington Post* (May 28, 1992) described the forebodings of a prospective groom who just "knew" the carefully orchestrated wedding plans would not materialize. On the day prior to the scheduled wedding, the bride-to-be and four family members were all injured in an automobile accident. In such personal examples and those of professional psychics, what is seen of the future are projected potentialities that cannot be changed *directly*, psychically or otherwise. Whether or not they materialize depends on intermediate actions undertaken by conscious choice or through an accident: In either case the interim events determine which potential event actually materializes.

In a universe of freedom for each individual to choose among different potentialities at each quantum step, no form of telesight, even with powers of retrovision and precognition, could include the power to revise past or future decisions made by individual beings in ordinary reality. While the mind can travel along the arrow of time, but not directly change phenomenal reality, telekinesis can influence the behavior of matter at a distance, but is by its nature limited to this space-time where matter resides. The fantasy of changing history through time travel may be generated by those with a need to escape responsibility for their own decisions, past or present. In this book's model, only through actions taken in quantum gaps—breaks between past and future permitting the introduction of uncertainty—can the web of cause and effect be evaded.

It is through taking advantage of such quantum leaps that conscious beings, in concert with others, become co-creators of future reality. However, the uncertainty principle, the play of pure randomness, still introduces a degree of chance into the evolution of ordinary reality, thereby restricting the degree of personal control. One of the conditions under which beings apparently agree to play in this universe is the risk of such accidental mutations. Conscious life carries with it these elements of chance, adding zest to existence: by remaining in a conscious incarnation, beings accept the challenge of riding the waves of creative uncertainty.

Other Beings

There exists still other evidence of a unified field of consciousness—other beings with whom humans can communicate through the subtle and physical senses.

Humans are not the only sentient beings on the Earth and apparently have never been. As previously mentioned, since the late 1940s, thousands of sightings of "alien" spacecraft have been reported around the world, with many involving direct or telepathic communications with the occupants. Physical interactions (including medical procedures and sexual contacts) have been verified through material evidence, hypnosis, and polygraph tests. The sources referenced in this book and many others provide credible descriptions of these activities.

That many individuals have been involuntarily subjected to alien experimentation is now well-documented. Ova have been taken from some women for fertilization elsewhere, while others have been impregnated and then had the fetus removed at a very early stage. Men have had sperm extracted or been forced to ejaculate into alien females. Some women have reported contact with their hybrid offspring.[15]

While shocking and outside the bounds of our concepts of civilized behavior, most of these encounters seem to involve no inherently hostile intent on the part of the alien abductors. They appear to have no desire to inflict harm and frequently promise further contact and communication. Their potential for empathizing with the psychological terror felt by some of their human abductees may be blocked by the nature of their own situation. Some abductees have received the impression that the aliens need the abductees' genetic stock and something of their emotional force to strengthen the aliens' own weakening species. Humans and some aliens may be mutually dependent in some as yet unrecognized way.

Evidence also exists of more threatening types of alien presence. As with humans, these beings are clearly at different levels of moral and psychological development. (See earlier referenced work by Sprinkle.) As disheartening as it may seem, having ended the need for taking sides in the Cold War, many believe it may be necessary in the near future to seek alliances with some aliens in defense against others. Currently, covert operations involving opposing alien forces of potential threat to human society may be underway. A forthright public assessment of the various species of nonhuman beings and their compatibility with society's most progressive values is needed. If this need is not accepted by people in formal positions of leadership, informal networks must evolve to take the initiative.

A number of widely reported polls confirm that large numbers of American adults are prepared to accept the reality of other beings, and the responsibility for dealing with them. Many claim to have had ET-related experiences themselves and, as these claims are taken more and more seriously, intelligent individuals will find ways to integrate alien actors into their own sense of reality. The three-faceted model of the universe postulated in this book will assist humans in developing plausible explanations for reported alien encounters and strategies for relating to them.

The first type of interaction occurs when alien beings are primarily operating in phenomenal reality. They are present in ordinary space-time and accessible through the five physical senses, even though they may possess characteristics and abilities normally unknown to us. This explanation accounts for physical traces left by ETs and their vehicles and reported artifacts. It could also account for the wounds or other marks reported on numbers of contactees.

The second category involves experiences with primarily imaginal forms that become perceptible to people in the requisite open mental state. Jacques Vallee, for example, hypothesizes that a significant element in the current reporting of UFO and ET contacts could result from interdimensional perceptions or access to alternate realities.[16] This theory explains the number of passive memories of alien contacts (obtained through the subtle senses from another party) that are exposed through dream recovery or hypnotic regression.

A third explanatory theory involves some nonhuman beings inhabiting the subtle energy realm, where they are encountered by people in altered states. Temporary shifts in their subtle energy vibratory pattern could explain why some individuals in a group can "see" an anomalous being while others cannot. Such a state of requisite "readiness" could also help explain the selective nature of contacts with beings that have been labeled "angels," "ghosts," "allies," or "nature spirits" by their human interlocutors.

It is not clear what mental or emotional states are involved in the last two categories of contacts. Frontier scientists can take the experience of dowsers who succeed in negotiating the departures of certain nonphysical beings as a starting point for research. One relevant theory, for which we have no direct evidence and which may be only a matter of semantics, provides for several dimensions within this space-time continuum. It may be possible for humans to shift from dimension to dimension along this unbroken spectrum without completely moving to another realm. Information reportedly from other beings supports the idea that there may be various dimensions within each of the three facets.[17]

A number of metaphysical books predicts humans are about to experience transition to higher dimensions or higher frequencies of vibration. Some of these allegedly involve states of consciousness different from those now experienced within the *phenomena*, but such speculation cannot be treated scientifically unless it can somehow be tested in this incarnation.

Extradimensional Assistance

A theme currently running through many groups (fundamentalists, new-agers, frontier scientists—even mainstream analysts studying population, health, and resource trends) is that human mental and psychological shifts of significant orders of magnitude are imminent. Some see these shifts as natural cycles, involving periods of ideational revolution and overt conflict, when declines in political, social, and economic institutions are paralleled by a widespread deterioration in the quality of life. Others think in more metaphysical terms, hoping for some form of divine intervention to accompany the arrival of a bright new millennium.

If cosmic consciousness is a unified field, populated with a variety of beings in different realms or dimensions, humans can benefit from inter-dimensional communications in a period of exceptional chaos. Much channeled material now being published seems to be helpful advice from more experienced races. The messages some alien abductors give abductees are perceived by them as helpful. Whether such communications can be beneficial or not ultimately depends on their validity and relevance to human perceptions of need.

Given the content of these communications, many believe ETs may be ready to assist humans in amending disastrous environmental practices or solving other pressing problems. Some see ETs as a catalyst to resolve internecine conflict among different political and ethnic groups. Conversely, others fear the imposition of a new—not necessarily benevolent—world order by covert human groups in concert with ET forces. The apocalypse, followed by a new heavenly reign on Earth, predicted by fundamentalists in various religious sects, is interpreted by some as the advent of an alien occupation.

The increasing number and complexity of unexplained "crop circles" in the 1990s have raised the intriguing possibility that humans are communicating with the consciousness of seemingly unrelated species, and even with Mother Earth and the Plant Kingdom. Some observers feel human behavior is eliciting responses from more evolved beings. Even after giving credit to hoaxers for creation of some of the formations, some researchers believe an interdimensional channel of communication is opening, waiting for systematic exploitation.

Even if nonhumans have no plans to mount an imminent public intervention in human affairs, humans still have access to broader knowledge. Regardless of alien intentions for the future, humans have sufficient latent wisdom and skill to bring about their own transformation. With the will to try, humans are now capable of species-wide shifts in consciousness through increased use of the subtle senses. Acquisition of vast new knowledge on the phenomenal plane can be facilitated by expanded understanding of the *energeia* and the *noumena*. Aided by current global communications networks, rapidly expanding human awareness would result in a radical transformation of basic assumptions about the nature and role of humans in the cosmos. A shift of this order of magnitude would be more than the acceptance of a new scientific paradigm. Willis Harman used the term "global mind change" to label the phenomenon of such a new perspective.[18]

While its full implication would take a few years to play out, as did the European Renaissance and it prehistoric analogs, this new "*perestroika* of the mind," would change the nature of all human institutions, and thereby the individual's relationship with others and the material universe. Such a revolution's ultimate pace and character will depend on the nature of leadership assumed by people who can now discern the emerging potentialities and highlight them for broader audiences.

Inner Power

The idea that certain men (rarely women) are gods incarnate has been perpetuated through recorded history. All the major Middle East-based religions have founders, and in some cases current leaders, who profess to be uniquely linked to the ultimate God. Eastern religions still profess to have their avatars and divine incarnations. The desire to believe that the "gods" are still present dies hard; clinging to this dream gives a sense of security to beings who have not yet recognized their own cosmic nature. Such beliefs also contribute to an assumption that extraterrestrial beings are in some way more "divine," rather than just cosmic beings different from humans.

The notion of divine/nondivine categories persists despite evidence to the contrary. But most people have heard at least one story about ordinary humans exhibiting special powers: religious men who exude the ash of *vibutti*, believers who ooze perfume from their hands, or others who manifest stigmata. The *Washington Post* has prominently described the case of a local priest—observed by credible witnesses—who causes statues to weep in his presence. The Catholic Church has a bureaucratic process for assessing such incidents and classifying those that meet certain criteria as "divine" miracles. Such miracles are few because the church has a vested interest in not acknowledging too frequently the special powers of common folks.

Unfortunately, religious thinking has separated humans from humans, interpreting descriptions of hierarchies in various cosmologies to mean different levels of godliness within humanity. European kings, and many others, until recently, continued to proclaim they ruled through "divine right." Another example is the misapplication of Hindu cosmology to justify a caste system that puts humans into separate and distinct spiritual categories. Such notions of "divine" hierarchies have permitted divisions based on prejudgments about race, culture, religion, education, and economic and technical development.

Inner illumination and scientific research recognize the universal nature of the human power to perform all the so-called "divine" miracles. Positioning themselves as somehow subservient to the "gods" narrows the scope of people's vision and intellectual inquiry. Seeing themselves as fallen souls or struggling animal-like creatures who seek deliverance by exterior gods, groups fail to exploit the cosmic reservoir of knowledge and wisdom all conscious beings possess. Having inadvertently convinced themselves that they are incapable of self-redemption, individuals leave social havoc in their wake.

The evidence indicates that all humans possess the inner power—though still latent in many—to deliberately transcend this fragmentation of feelings, attitudes, and behavior and achieve wholeness. They can express and perceive noumenal information in the far corners of the universe; view the past and the future; mobilize unseen energies; and cause matter to move and coalesce in tangible form. Yet these powers are not harnessed in the interest of harmony in today's society. The obvious question is, why do partially conscious beings deny the power within themselves? Is it fear or ignorance, or both?

Illusions and Delusions

As illustrated below, religious traditions the world over perpetuate the expectation of a return of a God who will usher in a new age of enlightenment or regeneration, thereby taking the responsibility from us. Nicholas Roerich wrote:

> *Kalmucks in Karashar are awaiting the coming manifestation of the chalice of Buddha.... On Altai, the Oyrots renounce Shamanism and are singing new chants to the Awaited White Burkhan.... The Mongols await the appearance of the Ruler of the World and prepare the Dukang of Shambhala.... The Jews await the Messiah at the Bridge.... The Moslems await the Muntazar... The Christians of Saint Thomas await the Great Advent.... The Hindus know the Kalki Avatar and the Chinese at New Year light the fires before the images of Gessar Khan, ruler of the world.*[19]

The historical analyses of Zecharia Sitchin and William Bramley[20] suggest that this sense of expectation may have evolved from the promises to return made by advanced beings as they left behind our human ancestors. Supporting this notion is the fact that the sixteenth-century Spanish conquistadors were welcomed by the natives of Central and South America whose legends predicted the return of powerful light beings. That behavior expressed the same yearning motivating the Jewish prophecy of a Messiah. If the basic tenet of a "divine" Messiah or Savior is based on a historical promise by aliens to return to Earth, what are the bases for other underpinnings of religion?

Current religions came from regional expressions of common human strivings for explanations of generic memories, for meaning in life, and for the meeting of similar emotional needs. The largely cosmetic theological differences that divide religions are much less important than the common needs and aspirations that bind humans.* By falsely attributing divinity to certain beings, the religions demean humans by failing to respect the individual's power and scope for participation in cosmic life. Religious leaders, vested in the current power structure and its personal benefits, have little motivation to stress the common ground that would heal social divisions and demonstrate underlying mutual interests.

Even though people can be held back by self-serving leaders of religion and other institutions, in the final analysis people demean themselves. When they turn to such authorities for spiritual answers, they give up personal responsibility. Some leaders may initially seek to inspire humanity to new heights, yet the limited expectations of the masses hold them back. A self-deprecating public forces the static collective reality on the would-be avatar who might otherwise manifest a more evolved state.

Millions of people, for example, believe that Sai Baba, a man from the village of Puttaparthy, India, is God Incarnate. Over the last 50 years, his miracles and teachings have attracted thousands of people for daily darshan (bestowing of a guru's blessing). Their time is spent in acts of adulation; by metaphorically grabbing for his feet they prevent a unified march toward their avowed common goal. Although Sai Baba says, "Look within yourself. We are all the same," his followers tenaciously deny the divine consciousness within themselves. While he promulgates a vision of equality among all beings, a code of community service, and values that support enlightened human behavior, the thousands awaiting his darshan hold strong expectations, from a magical cure for diabetes to a cost-free way to clean up the environment. They hope Sai Baba will accomplish these ends without any effort on their part.

Thinking of beings like Jesus and Sai Baba as gods incarnate, people ascribe to them the characteristics of all-knowing, all-seeing, all-powerful

---

* In his book *Sentics,* Manfred Clynes demonstrates that all peoples share the same basic repertory of emotions, the expression of which is measurable physiologically. These universal patterns are masked only by consciously controlled behaviors of various human cultures.

paragons, thus making it psychologically possible to accept a lower standard of behavior for themselves. Such chosen men and women are also set apart by the misguided view that they are rebels to be either crucified (Jesus) or urged into battle (Mohammed). Or, they are gods to be adored but not emulated (Sai Baba). When the respective devotees address their leaders as God, they avoid self-responsibility. Theistic purists conclude that these beings could not really be gods and, thus, permit themselves to ignore the message of unity and service. Both groups mistakenly presume an unbridgeable chasm exists between divine wisdom and human capacities. The result is a delusional escape from self-responsibility in the expectation of an external savior.

What is needed now are *metascience*-based myths that transcend politicized historical interpretations, e.g., the fourth-century Nicean Council censorship of the founding principles of Christianity, and reveal the commonalty joining humans and other species. Such a Solarian myth will reflect frontier science's understanding that the universe and all its beings are spiritually and materially undivided and indivisible. The true "masters" of the twenty-first century will not be specialists who devote all their attention to mutually exclusive religious tenets or the self-limiting restrictions of current science. Only a timely synthesis of inner *and* outer wisdom can spare the present civilization the apparent self-destruction that has befallen those in the past.

Self-Responsibility

Communications with more advanced beings will help humans realize more fully the scope of their own powers. When humans psychologically accept as siblings other conscious beings with powers that have traditionally been labeled "divine," they will in effect be accepting humanity's cosmic nature. Part of the process of assuming the mantle of cosmic self-governance must be a deliberate search for unknown family members. Humans must deliberately try to communicate with cosmic siblings through the subtle senses, as well as through the channels of the physical senses. *All* senses and instruments need to be aimed and tuned in the right directions.

On October 12, 1992 (500 years after Columbus "discovered" America), NASA launched the 100-million-dollar SETI (Search for ExtraTerrestrial Intelligence) Project to listen for possible radio signals from intelligent extraterrestrial beings. This program was justified on the statistical assumption that humans might not be alone in the universe.[21] If there are intelligent beings, so the argument went, they might be trying to communicate with us, and therefore we ought to electronically scan the skies for messages.

The project involved astronomers using telescopes in Puerto Rico, West Virginia, California, and Australia, and computer centers around the world. Designed to scan the quietest sector of the radio spectrum (1,000 to 10,000 megahertz), SETI sought—among the billions of frequencies—signals that stand out from natural noises and emissions from Earth. The logic went like this: Intelligent beings would use one wave length, keep it on for a period of time, call attention to it by transmitting in pulses, and compensate for the Doppler effect of their moving planet.

Fortunately, that expensive project has been defunded by Congress. A small NASA radio telescope has been launched to carry out part of the SETI search outside the distortion of Earth's atmosphere. Few public resources should be devoted to a low-priority task when so much concrete evidence of nonhuman consciousness exists here on Earth. To ignore the evidence is analogous to staring at the sky for a single source of music while ignoring the song birds in the trees. Society can no longer afford to ignore evidence in its front yard out of either intellectual conceit or fear of public reaction to the truth.

Carl Sagan, whom many considered to be the government's informal debunker of alien research, writing in the mass-appeal medium of *Parade Magazine,* appeared in an unintended way to be preparing the public for the inevitable alien encounter. In September, 1993, he wrote of SETI, "Conceivably, this might be the last generation before contact is made—and the last moment before we discover that someone in the darkness is calling out to us." He attempted to be reassuring when pointing out SETI was passive and not likely to be detected by malevolent aliens. He reminded readers that since current science and knowledge will soon be outmoded by human progress, more advanced civilizations should not be feared. Recognizing the likely widespread impact of the denouement of ill-founded societal assumptions, before his death, he implicitly cautioned that acceptance of proof that humans are not the universe's only conscious beings will dramatically reshape politics, ethics, economics, and religions.

Individuals who have the most vested in the political and economic status quo may consider this inevitable revolution to be reason enough for the continued public denial of the growing evidence of UFO/ET activity. In 1960, the Brookings Institution in Washington, D.C., prepared a report for the U.S. Government which warned that public knowledge of the existence of other intelligent life in the universe might lead to societal disintegration. Many believe fear of such an outcome has motivated almost 50 years of government cover-up.

Such fears are ill-founded: ordinary human beings, as is so often the case, are way out in front of their formal leaders. Many already recognize that opening to cosmic consciousness is essential for planetary progress; they understand that true participation in cosmic life is a collective experience, shared not only with fellow humans, but with all conscious beings.

To consciously claim a species' place in a singular or integral universe requires recognition that beings from all levels—angels, devas, spirit guides, and aliens—need each other. None exist just to give instruction or issue commands; their communications represent only their own experiences, from which others can learn. Through such egalitarian and symbiotic exchanges, the species can nourish each other. Even those with hostile agendas must be dealt with in the same manner.

Currently on Earth, crime and violence anywhere constrain freedom of movement in the neighborhood or around the globe. Pollutants in the atmosphere deplete everyone's ozone. Poor ecological practices deprive all of wholesome food, air, and water. Emotional and mental alienation in some weakens the fabric of a whole species. Children are discarded, and even murdered, as adults fight over ephemeral issues and cringe with inner fear of honest and constructive engagement with other beings.

The result is species-wide alienation from its true nature in which each member is held in a lower, more unstable path by the energetic forces that influence cosmic positioning. If all creative energy is spent in avoiding responsibility for oneself, the web of defenses does not change and perpetual fears remain in place. But when some individual energy is spent helping the whole community progress, and in giving more physical, social, and psychological freedom to other beings, all are individually freer.

Why do humans attempt to avoid responsibility for discovering the truth about themselves? The highly educated, affluent devotees of a channeled being, observed by this writer, displayed an easy acceptance of channeled platitudes, while revealing a deep need for reassurance that they were his special charges. Perhaps this need to remain dependent in a "lower status" is a function of the anxiety felt when sensing one's own latent powers. Humans may be frightened by their own potential for greatness.

One way people can avoid the myopia of devoteeism is to remember that they share the cosmic nature of the inappropriately elevated person (or being). Sai Baba, for example, admits to accepting the Hindu tradition of dividing men from women in public, even though it contradicts his teachings. Just over five feet tall, he compensates for his short stature with wildly blown hair, exaggerating his claim to omniscience perhaps to bolster the confidence of the village child he once was. Such are the psychological defenses of charismatic leaders—in material or energetic form—who either forget, or are not allowed, their natural limitations.

If human society is to perform at its peak, with everyone moving higher on the cosmic spiral, the false distinctions that set leaders and followers apart must be abandoned: all beings must relate to other cosmic beings as equals. That is the privilege and responsibility inherent in the Solarian Legacy for human society's twenty-first century.

NOTES

1. O'Leary, Brian. *Second Coming of Science*.
2. Steiner, Rudolf. *Cosmic Memory*.
3. Prior to Rupert Sheldrake, the idea of a morphogenetic field was advanced by Paul Weiss in the 1930s and more fully developed by J.V. Bronsted in the 1950s.
4. Wilhelm, Richard, and Cary F. Baynes. *The I Ching or Book of Changes* (Princeton University Press: Princeton, New Jersey, 1975).
5. Hannon, Robert J. "Letter to Editor." *Mensa Bulletin*. Jan.-Feb. 1994.
6. Sitchin, Zecharia. *Earth Chronicles* and *Genesis Revisited* (Avon Books: New York, 1990).
7. Copies of papers may be obtained from IANS, 1304 S. College Ave., Fort Collins, Colorado. 80524, USA.

8. Kean, Linda. *John Lennon in Heaven* (Pan Publishing: Ashland, Oregon. 1994).
9. Route 1, Box 175, Faber, Virginia 22938; (804) 361-1252. The Institute is about 25 miles south of Charlottesville.
10. Borbely, Alexander. *Secrets of Sleep* (Basic Books: New York, 1986).
11. Stevenson, Ian. "Reincarnation: Field Studies and Theoretical Issues." *Handbook of Parapsychology*. Editor. B.B. Wolman (Van Nostrand and Reinhold Co.: New York, 1977) and Keynote Speech. IANS Conference, Fort Collins, Colorado, 17 Sep. 1992.
12. Moody, Raymond A. *Life After Life* (Bantam Books: New York, 1988).
13. Ring, Kenneth. *Omega Project* (Morrow: New York, 1992).
14. Fowler, Raymond E. *The Watchers: The Secret Design Behind UFO Abduction* (Bantam Books: New York, 1990).
15. Hopkins, Budd. *Intruders* (Ballantine Books: New York, 1988); Mack, John E. *Abduction.*
16. Vallee, Jacques. *Revelation: Alien Contact and Human Deception* (Ballantine: New York, 1992).
17. Alder, Vera Stanley. *The Fifth Dimension: The Future of Mankind* (Samuel Weiser: New York, 1970). Many others have followed over the past quarter century.
18. Harman, Willis. *Global Mind Change* (Knowledge Systems, Inc.: Indianapolis, Indiana, 1988).
19. Roerich, Nicolas. *Archer* (Society of Friends of Roerich Museum: New York,1929).
20. Bramley, William. *Gods of Eden.*
21. Assume there are about 100 billion galaxies, each with billions of stars. If only 10 percent of those stars have the life-giving capability of our sun, and only 10 percent of those life-giving suns have planets similar to Earth, statistically there could still be thousands of other advanced civilizations.

Chapter 6

Result: Local Incarnations

The stars and planets of a grand universe abound, but modern technology has probed the depths of its microcosm to discover reality is ephemeral. Between these extremes, suspended in space, is Earth, home base for many conscious beings. The experienced reality of all these beings encompasses the mind, body, and spirit as facets of the whole. As cosmic beings, they derive power from and thrive in the interactions of all three realms. They perceive cosmic reality through multilevel sensory systems that also help shape it, in a circular process of co-creation. Who created the concept of a being, and the force to bond it with matter? How is conscious life breathed into inert matenergy?

◆ ◆ ◆

I FIRST HEARD THE WORD "*beingness*" while listening to a taped psychic reading done by Ron Scolastico[1] in 1979. He was in Iowa, I was in Washington, D.C. A mutual friend had asked Ron to go into a trance and seek his spirit guides' responses to questions I had written.

The spirit guides, speaking through the channel that Ron allows himself to become, used a very stilted syntax, not unlike some of my early efforts to translate English thoughts into French or Spanish. One of the first things to strike me in that recording was the reference to my current life as "this *beingness*." As their comments unfolded it became clear that this was a good use of English for the guides' meaning. By *beingness* they implied my specific incarnation in space-time. The implication was that this *beingness* was me, but not all of me; the *beingness* was bounded in ways that I was not. I much later came to understand that *beingness is the organism manifested when patterns of consciousness are focused in subtle energy and* matenergy. In traditional terms, *beingness* is equivalent to incarnation, where consciousness that already exists takes on a different form. Therefore all entities with conscious awareness, regardless of species, are incarnated beings. Some believe the Earth itself in this context is a conscious *beingness.*

Universal Consciousness

Where does the concept of *beingness* or incarnation fit in the context of frontier science and the three-faceted model? An incarnated being at its primal level arises from what David Bohm called the "implicate order" and Paul Tillich called the "ground of being." It starts as a tiny, but perceptible wrinkle in the continuous, yet variegated plasticity of cosmic consciousness, i.e., a pattern within the *noumena.* Over time it gives rise to small pulsations within the cogitating membrane, emerging in relation to similar wrinkles jostling for position in a field of subtle energy filled with intersecting patterns of intent. Although clothed in this field and embodied in *matenergy*, there is not a solid membrane dividing one *beingness* from all others.

The individual body is but a delicate pattern of cosmic position-holding, analogous to the air space of one plane among many others circling in their various landing approaches. These holding patterns of *beingness* are maintained by clusters of intentions not unlike the inner images of a performer on a tightrope, requiring the maintenance of place and form without permanent structures. In this metascientific context, a being, in sum, is a set of bits of matter (itself only quanta of gross energy) focused by conscious intent in a field of subtle energy.

There is currently very little comprehension about that tenuous link between consciousness in a being and general consciousness. If one assumes the human experience of *beingness* is not unlike that of all species and dimensions of *beingness*, it can be inferred that the following descriptions are relevant to all conscious beings.

Human experience indicates that during life individual consciousness never totally breaks off from the group mind (cosmic consciousness). As a being defocuses (in death) from material bonds, it may meld back into cosmic consciousness, yet retain some degree of the integrity experienced during incarnation. Given the evidence we have that some departed personalities are able to continue to interact with the living, it appears that an individual identity can persist for an indefinite period beyond death. This tangible evidence is the strongest case that can be made for a form of immortality, believed by some to exist in the "noosphere," perhaps a combination of the *energeia* and *noumema*.

When, within this incarnation, there seems to be little or no volitional control of the boundaries of consciousness, human psychologists label the individual "psychotic." If the boundaries are intentionally permeable, the being is called a mystic, meriting respect and emulation. Both so-called psychotics and mystics generally define their respective states in relative isolation from others. When there is the need to engage socially, the psychotics are usually coerced by institutions to give up their "insane" communications. If psychotics resist, they are drugged to insensitivity. A more humane approach would use "subtly sensitive support" from others, involving the reestablishment of joint parameters to define self (see fuller discussion in Chapter 8).

Both the mystic and the psychotic teach us that maintaining the integrity (both mentally and physically) of any being requires developing a certain degree of congruence with the expectations of the larger community. The term "joint parameters" is appropriate: a being must initiate the creation of self, but must work together with others at the same time, taking into account their acts and thoughts. Ongoing personality development is a collective function, involving the embeddedness of the individual in the whole through the energeial and noumenal connections.

One can deliberately connect with the realms of other beings and objects, including species and collective memories (see the work of Stanislav Grof[2]) through the porous field of consciousness. Without falling into a mystical or psychotic state, one can give oneself permission to range far and wide, in the manner of a shaman[3] who walks daily with one foot in each world. One can choose to expand the scope of conscious experience far beyond ordinary definitions of personality.[4]

Moving with ease in the sea of consciousness, surfing the noumenal cosmic waves, one can engage in the play of "synchronicity." The phenomenon, so labeled by Carl Jung, manifests where physical events seem to conspire to meet one's needs for information and interaction with others. The occurrence of synchronous events in a life is correlated with the degree of conscious openness exhibited.[5] The implicate order supports and accommodates purposefulness. People get what they express a need for through prayer, holding a vision or clear intent, and positive thinking. Understanding the workings of synchronicity at the level of individual being provides insight into the functioning of the universe as a whole.

As discussed earlier, the Principles of Mentalism and Correspondence insure thoughts affect *all* realms. When a thought is "consciously" expressed, it reverberates in the subtle energy field. Like the proverbial stone dropped into a pond, it moves out in ever-widening circles, but in *all* directions (not just the two-dimensional water surface). The vibrations of a thought pattern gather to it subtle energies involving the relevant forces, giving rise to powerful forms in the *energeia*.

The resulting thought form, in its energized state, interacts with organic and inorganic systems, affecting people, plants, animals, and machines. Radiating and becoming mingled with other forms and beings on the same frequency, it creates new energy flows and events in the never-ending process of creation. While intentions may be absorbed by more powerful forces of a similar nature, or may be largely neutralized by the opposite polarity, an impact discernible by the physical and subtle senses is always inevitable. The admonition "don't wish for it unless you really want it" is sound advice.

The quality of life for any *beingness* is therefore a function of the quality of the consciousness it brings into space-time. When thoughts remain negative too long, cells are thrown out of balance and become diseased. Animals and plants are agitated, and materials in machines and tools exhibit stress. Positive thought forms have the opposite influence. But neither radiates unimpeded throughout the universe; each one is poised, ready to fit into a receptive niche—an open being, an unprotected system, or neutral matter—that reciprocally shapes its materialization.

Whether the thought's vibrations penetrate the thinking of another depends on its strength and precision, and also on the state of the potential recipient. Both parties must be "in tune." The intended receiver can be either open or self-defended. When beings are relatively unprotected they are susceptible to vibrations consciously focused in their direction. Experience shows that thoughts of hatred and alienation from a few can infect a society. (There are reports of humans being so unprotected that they are taken over by other beings called "walk-ins."[6]) Being open is not the same as being unprotected: the difference is in degree of conscious management of one's boundaries. Works by Richard Gerber and Barbara Hand Clow offer insight into the human ability to play openly with the flow of energeial and noumenal forces while protecting one's own development for more effective interactions with others.[7]

Experiments with focused telepathic communications are so successful—both in sending and receiving—that it does not seem farfetched to conclude that each being possesses a unique subtle energy signature. Though the process by which such communications work has not been discovered, people can focus on a "target," using personal identifiers or abstract location, and send and receive images or thought forms. (Perhaps the Internet is a mechanized metaphor for this phenomenon.) These transmissions can be a combination of an abstract idea (word, concept, or form) and an emotional charge (fear, awe, anger, love, reverence, etc.).

When telepathically receiving the vibration of thought forms, most people are hazy about the ideational content. They project interpretations onto the other much in the same manner that dreamers apply their own content to externally generated stimuli. The emotional loads in the messages—because they are generic in humans—are generally more easily understood, even across cultural lines. The research of Manfred Clynes demonstrates similar, measurable, internally-generated patterns of physical tension that arise when different individuals experience the same emotions.*

* Measured by sentographs, these recorded patterns of pressure exerted in a finger are "sentic forms" which correlate with basic and distinct emotions. [8]

In the nonhuman context, certain thought forms are believed to exist as separate "elementals." They have a limited life of their own, capable of influencing natural processes, as in gardens. Other thought forms may be used to intervene in natural phenomena such as weather patterns and behaviors of groups of plants and animals, and even larger systems.

Self-Definition

If consciousness is indivisible, as manifested in split-particle physics, cell-to-cell communication, and ESP, what is an appropriate conception for the individually incarnated particle of the cosmic soul or mind?—only a cell in a larger organism?—a fragment of a cosmic hologram?—a dream image of Ultimate Consciousness itself?—a lower level of consciousness? To hold steadfastly to any particular formulation is currently an act of hubris. The most we can do is make some educated extrapolations—by synthesizing the evidence from various fields of inquiry and experience—and create a limited concept of cosmic *beingness*.

The biological determinist would stipulate that it is the patterns of DNA that program development from zygote to adult. The priest would argue that it is a function of the soul incarnated at conception. The behaviorist would assert that a being is conceived as a *tabula rasa*, with its character shaped entirely by the natural and social environments. We have recent evidence (discussed in Chapter 2) from the studies of biology and consciousness that indicate development is a reciprocal process. Mind is at the center, controlling the process, but also taking into account feedback from the biological organism. This perspective has many implications for new hypotheses. The reality of statistical odds seems to be on the side of those who believe some patterns (forms of consciousness) preexist human conception and are required to inform the act of conception itself.

If the conscious pattern precedes the cohesion of the energy body and the formation of the physical body, the termination of a foetus is no different than killing a 12-year old or a person a half-century in age. In all cases, a particular conscious incarnation is irrevocably destroyed. If the essence of the being is fused with its physical host from the moment of conception until the breath of life leaves the body, the age of the physical carcass in no way diminishes the inherent integrity of the "soul."

Current human genome research relating to homosexuality and schizophrenia leads some to believe that genes influence a being's behavioral pattern, as they appear to determine physical attributes, thereby adding fuel to the long-standing heredity-versus-environment argument. The determinists seek a particular gene or cluster of genes relating to a personality trait (schizophrenia) or behavioral predisposition (homosexuality). When they find a certain degree of correlation, they speculate that various profiles of gene maps interact to shape behavior. Then they infer causation.

To stipulate that a biologically fixed gene pattern precedes any act of consciousness denies the possibility of nonlocal* existence and reincarnation. On the other hand, accepting the evidence that patterns of energy (behavior of parents) and forms of matter (gene positioning) follow—within certain cosmic principles—conscious intent, makes conceivable a rational/scientific explanation for a process of conscious incarnation. The resulting genetic patterns would provide the mind (interdependent consciousness) a certain keyboard on which to play, where the selection of the tune is a conscious act. If this concept of incarnation is valid, it is logical to conclude the focus of consciousness expressed through conception holds the being together through its physical cycle.** Beings with enough individual consciousness development may be able to maintain the integrity of a Self through successive incarnations. The stronger the mental patterns over one or more lives of growth and self-direction, the more likely the perpetuation of an individual integrity through the rigors of death, birth, and childhood in the process of successive reincarnations.

Regardless of the validity of any particular transpersonal hypotheses, each being is faced with the daily challenge of defining oneself in relation to other beings and the material world. Since a planetary being has at least three facets—physical body, subtle energy field, and mind—maintaining the integrity of self is a multilevel process, ranging from largely instinctive

---

* The use of the terms "nonlocal" and "local" by physicists, and in this book, is itself not fully free of dualistic-thinking. We need a term that encompasses both.

** An interesting challenge will be to decipher the manner in which cloning (creating duplicate plants and animals, and eventually humans, from the DNA of an existing being) affects mind-body interactions.

physical acts of self-nourishment to highly structured regimens of exercise, diet, and conscious engagement. The more consciously aware realize the need for individualized and progressive control of their energy patterns and behaviors. (Although self-determination is inherent in human life, it is easier for people to go along with socially accepted physical and social norms. Traditionally defined jobs, professions, and other roles help an individual shortcut the process of self-definition.)

Just as some beings' bodies are more resistant to viral or bacterial invasions, so some minds are more resistant to penetration. Resistance on the physical level, including the maintenance of a healthy immune system, has its analog in the permeable membrane of consciousness. Susceptibility to spirit attachment, possession, multiple personality, or delusions is a function of one's psychological immune system, the personal subtle energy field in the *energeia*.

Because of the less obvious nature of subtle energy, conscious attention is required to maintain personal integrity at that level. Most humans are conscious of the need to monitor their feeling states and take appropriate actions to keep on a positive track. They realize some degree of self-direction is not only possible but desirable. They know enlarging the sphere of self-direction frequently requires extending one's personal reach and attempting to change societal patterns.

Beingness, as the local manifestation of a given cosmic being, never stops changing. The Principles of Vibration and Rhythm result in its being impelled by the inertial force of accumulated experience, and simultaneously drawn by a vision of what might be. As in composing a symphony, the being has an almost infinite variety of motifs from which to choose. However, after completing the score, the being must depend on other musicians for its performance.

Community Support

Complementing the being's self-definition, a necessary basis for its integrity, is the collective degree of support or opposition the being receives from its primary community. (See the Chapter 8 for further discussion of these interactions between Self and Other.) The constraints of the larger community and the natural world come into play with individual intent in the evolving character of a *beingness*. For example, the more other members of a community permit the acting out of individuals' constructive or benign intent, the freer they become to explore their own potentials. Freeing others frees one's own creative energies. The more supportive beings are of others, the more support they receive. This principle also describes the nature of the interactive cosmos on nonpersonal levels, alluded to earlier in the introduction of synchronicity.

In an interactive universe the negative actions (as well as positive ones) of all beings come back to haunt them. Even though, when a course of events is set in motion, one can never predict the final outcome, its effect will return to the initiator. The Hermetic Principle of Cause and Effect—reflected in the Hindu concept of karma and in the Biblical injunction "as we sow, so shall we reap"—works on currently unimagined levels through the interconnectedness and indivisibility of consciousness.

As intent is the most important element of any action in determining the results of individual behavior, so it is for the collective. Why is that so? As beings physically communicate, their inner message speaks telepathically more strongly than either voice or actions. Other cosmic beings perceive it through common access to the *energeia* and *noumena*. It is therefore not surprising that hostile intent evokes a defensive response from the core of the other.

One may realize short-term gains by trying to conceal true intent, but the chain of actions and reactions sets into motion a cycle of reciprocity on the inner level. Beings who have been originally deceived will not continue to be, and ultimately their lack of support through later connections, direct or indirect, will derail the progress of one's plans. Repetition of negative patterns can also debilitate the actor's own spiritual and psychological well-being. (The case of someone like Lee Atwater—the former Republican "dirty tricks" strategist who died of a brain tumor—is illustrative.) This cosmic principle is reflected in the Native American warning that the sun should never set before a hostile arrow is recovered and wounds are healed.

Limitations

Notwithstanding the power of intention, there are constraints beyond the social on the scope of a being's creativity: they result from the interdependence of physicality and consciousness. How much does the nature of a physical incarnation in this dimension limit the scope of conscious play? In-born influences on human capacities have a role in this context. Current IQ and behavioral research on twins enables us to make inferences about the interactive process and to formulate new hypotheses.

Many mainstream researchers start with the assumption that genes influence 70 percent of an individual's behavior, setting at 70/30 the "nature/nurture" influence ratio. However, their studies have the implicit objective of "proving" that genes, either those internal to the individual or the genes in others that influence him or her, determine 100% of behavior. This assumption becomes evident when one takes into account the scientists' assertion that the others in the environment are also genetically controlled. The inescapable conclusion is that some form of genetic influence determines 100% of behavior. The possibility that non-time-bound consciousness is an important influence in both the physical and social realms necessitates reinterpretation of the data and rethinking of these conclusions. From this perspective, testing the hypothesis that the 70/30 ratio indicates the scope a cosmic being has for self-determination in an Earthly incarnation could yield instructive results.

About seventy percent of all foetuses appear to start out as twins, but in most cases one atrophies early in pregnancy. Of actual twin births, about one-third are identical (created out of split halves of a single fertilized egg). Empirical data on the experience of identical twins, and to a lesser degree, fraternal twins, offer some intriguing possibilities for insight into the power of local consciousness to shape physical factors and behavior.

Identical twins report uncanny similarities in lifestyle and even in the making of specific decisions. Likes and dislikes are frequently parallel. They say they feel each other's physical pain and often telepathically receive each other's thoughts. They sense a strong emotional bond and suffer profoundly when it is disturbed. At the same time, they experience a strong impulse to individuality which frequently tests the limits of interpersonal tolerance and understanding.

Twins are acutely aware of the ties that bind and more easily recognize the effect of disturbances in such ties. They intuitively perceive that their individuality is a function of both interdependence and independence. In other words, in demonstrating the reality of the interconnectedness of all beings, twins are just like the rest of us, only more so. The chaos perceived by separated twins is simply an exacerbation of the disorientation everyone feels in a society characterized by alienation.

The local inflow of consciousness in the twofold conception* in one egg is less differentiated than in single incarnations. At the inception of foetal growth each twin is therefore as identical as can be, given the minute differences in matter that is concentrated by divided (but closely related) consciousness.**

Twins experience sharper ESP than the rest of us because they are more in tune with each other, having come from the same preincarnation field of consciousness and being endowed with almost identical physical sense receptors. They start out in almost perfect attunement, illustrating the impact of the preconception imprint on a being, and become less attuned as a function of individuation.

Fraternal twins experience a lesser degree of synchronicity than identical twins because they are "informed" by less closely related consciousness and by only half as much identical material imprinting. Therefore their degree of physical commonality would fall in the 30 to 50 percent range instead of the 70 percent common influence felt in identical twins, and their mental and subtle energy attunement would also be less. Current research on the development of siblings (twin and otherwise), rather than proving that matter unilaterally impacts thought and behavior, may indicate the opposite: the influence of consciousness on matter and physical traits.

---

* The experience many parents have of being joined by another conscious being at conception and during pregnancy reinforces the idea that spirit and soul incarnation occurs at conception.

** The experience of searching for a "soul mate" is possibly based in some such preincarnation connection that is severed by separate births.

Phenomenal Chance

Another characteristic all beings inhabiting the universe share is being subject to its built-in "software" operating systems. One of the attributes of the phenomenal realm is the statistical probability of flaws being introduced at the level of quantum gaps. An accidental flaw can appear in the physical manifestation of consciousness when DNA strands slip into an incorrect sequence; the delicate chemical balance of certain molecules is jostled; a neural link is severed; or an accident occurs in the melding of energy forces. It is as though the clay through which the cosmic artist works has an imperfection in the mixture: the work of art has a minor flaw, although the conscious design was true.

In humans, such accidental deviations from the intended pattern may result in epileptic seizures, Down syndrome, misshapen limbs, defective organs, etc., but they do not detract from the beauty and perfection of the conscious being. The affected being sends well-formed ideas into the *energeia*, but the message gets distorted in the physical transmission. Through help that keeps the communication clear, provides artificial aid, or overcomes the physical blockage through implants or stimulation of neuron regeneration, the full scope of consciousness can shine through. The life of Stephen Hawking, whose long battle with a motor neuron disease has not reduced the clarity of his acumen as a physicist, is a magnificent model of how such a quantum flaw can be bypassed.

Gender

Another key attribute of *beingness* is gender. The Principle of Polarity, operative at the subatomic levels of positive and negative charges, also manifests itself in the sexual nature of the universe's co-creative process. Every being has both masculine and feminine characteristics (Principle of Gender). Femininity is the state of receptiveness, involving openness and the nurturing of seeds from masculine expressiveness. Following the Principle of Rhythm, these two characteristics, extending beyond one's physical sex, tend to balance out in each individual over time.

For the female, the act of giving birth is masculine, requiring great force and concentration, following the feminine period of incubating the new being. Motherhood and fatherhood for the newborn are not naturally distinguished along sexual lines: both parent roles involve masculine and feminine aspects. Gender patterns are grounded in cosmic (Hermetic) principles, but the specifics of social sexual roles are largely a function of local traditions.*

The balancing of gender over a lifetime has interesting external manifestations. It is not uncommon for a man or woman who has been rigidly heterosexual to have homosexual experiences or relationships later in life. People who have been the most domineering in early life appear to need to become more dependent in later years. Partners who enjoy gender and sex-role reversal throughout their lives seem to enjoy greater inner balance in their old age.

Research into the causes of human homosexual behavior has resulted in speculations about social and biological influences that are mutually contradictory. The range of actual homosexual and bisexual experiences, particularly those of transsexuals, indicates that sexual self-identification is not exclusively rooted in the biology of birth, but neither can it always be correlated to social factors. These disconcerting findings call into question assumptions about the primacy of either biology or environment. The primary factor in both sexual and gender behavior may be the phase of transcendent consciousness present in the individual being at a given moment.

One of the tragedies of Western civilization's misunderstanding of the distinction between gender and physical sexuality is the romantic myth that each person must locate and forever bond with his or her mystical, opposite-sex other half. The myth causes an individual to believe the gap can only be filled by a single person, but social union alone, as in the convention of marriage, proves to be insufficient. Lifetime equilibrium in gender balancing occurs in the inner being, not as a result of the acts of another. A friend, colleague, or lover can provide only the raw material feeding into the internal balancing of gender within the individual. Therefore, the choice of sexual behaviors in this seeking of balance can be understood only in terms of the individual's internal rhythms.

---

* As late as the 1950s, nineteenth-century role assumptions were accepted as cosmically ordained. The next decades shattered such bland notions; the pendulum swung in extremes for both sexes. Thoughtful people have now realized a healthy community of cosmic beings requires gender balance and harmony.

Permeable Membranes

Neuroscientists who believe thoughts arise from neurons say individuals are literally alone with their thoughts. Nothing could be further from the truth. The undifferentiated and interrelated nature of the *phenomena, energeia,* and *noumena* intersecting in a *beingness* means each entity's boundaries* are highly permeable. The reciprocal flow through these membranes between a being and its cosmic setting alternates between positive and negative (Principle of Polarity) and expressing and receiving (Gender), both following the inherent Principle of Rhythm that requires rotation.

Fortunately, not all details of the interactions between a being and its environment have to be monitored in a focused way. Eyelids, sensing an impending invasion of sand, close automatically. Drivers develop a "feel" for the fenders of their cars while parking. Learning to drive, as in learning a sport, involves the heightened state of alertness that comes from incorporation of the subtle senses. Their unbidden signals daily protect us from error, accident, missed appointments, and losses, big and small.

Most people have experienced gaps in their focused awareness being filled from a different level of consciousness.** Taking full advantage of the process involves the maintenance of a flexible focus that is permeable enough to receive subliminal input. Blocked or congested five-sense information flows are covered by phrases like, "fear rose in my throat" or "his jealousy blinds him to...." Significant distortions of perception and judgment quickly follow if similar blockages occur in the subtle senses. Clearing the impediments from distressed or conditioned membranes must be a conscious process, and practice improves the success rate. When intentions and practices are harmonized, contributions of subtle-level communications keep people better integrated.

---

* Boundary issues are a topic of discussion in the current popular literature of psychology, but this chapter focuses on dimensions not covered in conventional psychology-subtle energies and fields of consciousness.

** While drafting this chapter, involved in a minor neighborhood clash over zoning standards, I dashed off a broadside of a letter. Although my letter was rational, something seemed vaguely amiss and I felt pulled to call the city office. Forewarned by intuition of the hole in my case, I hastened to deliver an amended note. My permeable membrane had protected my integrity, where consciousness either already knows (clairvoyance) of the missing information or, as a result of mental struggles, sends out a covert scouting party (remote viewing).

Times colored by fear-based religion, reductionist science, and deadening economic exigencies foster the closing down of the permeable boundaries of consciousness. The current collective fear of death (ironically, demonstrated by the brutality with which people, directly or through surrogates, mete it out) signifies fear of anything that would move today's human closer to the boundless realm of mind. When humans fearfully avoid any form of mind games, sex, drugs, physical play, or rituals that threaten to break through the barriers to the unknown, they cannot experience the expansion of consciousness that offers a higher level of development. However, the two poles require balance: openness must be countered with an appropriate degree of self-defense.

Self-Protection

An understanding of the need (discussed earlier) for balancing openness with self-defense comes from the world of Voodoo. Stories tell of people being bewitched by others, falling prey to the hexes of witch doctors, or succumbing to the spell of another's incantations. The way to avoid becoming the victim of another's intent is warding it off with one's own magic. Finding the power to protect oneself from the evil designs of another may require assistance from a friendly sorcerer or from a particular ritual that girds one's psychic defenses. But ultimately, regardless of the procedure used, the self-expression of intent is the source of defensive power. Only the being can deploy his or her power into mental space. For example, the Kahunas of Hawaii teach people to use their own inner resources, either explicitly or implicitly, to cast a perimeter of safety around their *beingness*.

Scientific evidence for the existence of the ability to cast some sort of protective shield around oneself may be coming from an unlikely source: a commercial piece of hardware that reads electronic signals. A Japanese company, Fujitsu, Ltd., is developing a computer that distinguishes "yes" and "no" thoughts by reading brain waves. If people can send messages that register electronically, they can also transmit energetic or noumenal messages that reach other beings.

The process of individuation, or setting up boundaries as a conscious being, has implications throughout the incarnation. In establishing its boundary, the being is erecting a protective membrane of conscious intent that will retain its integrity as long as it is mentally attended. The operation of this dynamic, for example, explains why no one can be hypnotized without giving concurrence. To be hypnotized, people must willingly allow another to penetrate their *beingness* and extract material from their pool of consciousness.

Honoring the multilevel flow of communications between self and others, yet avoiding being overwhelmed, is a delicate task. It is easier to ward off unwanted intrusions in the physical realm than at the mental and subtle energy levels. However, all restrictions placed at these levels, as in the physical, have a cost.

In the physical plane, humans can wear surgical masks, sterilize hands, or wear condoms to avoid exchange of harmful matter. On the plane of subtle energy, they can keep distant from one who wants to stifle or harm them. They can suppress their own emotional outbursts, and decline to invite people into their homes who remind them of their own pain. On the mental plane, humans can deaden their awareness through excessive consumption or through drugs. They can deny messages that come through the subtle senses. They can close their minds to new thoughts and ideas, weakening emerging fields of consensus being developed by others.

Ironically, more concentration of consciousness and energy are required to isolate the self one seeks to protect than it does to deal with and transform the interactions one tries to avoid. Such head-in-the-sand reactions in business, politics, and intimate relationships are commonplace. They are based on the ill-founded assumption that out-of-sight, out-of-mind tactics keep one from being tainted, but the inner linkages cannot be escaped. Paradoxically, the more people accept the reciprocal connections among beings and ride their ebb and flow, the more individuals can ultimately shape themselves in their own fashion.

Boundary maintenance is necessary to health and wholeness and involves interactions from routine social contacts to encounters with alien beings. Kenneth Ring believes "boundary problem people" appear to be more susceptible to negative UFO abduction experiences and suffering from overt physical or sexual abuse. Some therapists have concluded that helping people "shore up their boundaries"[9] is the way to stop such experiences or gain some control over them. The literature on abduction by alien beings indicates some people regard their experience as fear-filled while others see such experiences as growth-producing. The difference is often not in the experience but in the abductee's definition of self, which, as we have seen, is in the final analysis a largely conscious act.

Conscious Interaction

Beingness requires a continual process of incorporating sustenance from the cosmic habitat—in the forms of food, emotional energy, and ideas. The choice of the element for ingestion in all three categories defines the degree of conscious awareness in the being: the more conditioned and automatic the being's behavior, the less conscious and free it is.

Within the universe's built-in software, the impulse to self-realization seems to be innate. (Recall the concept of entelechy introduced earlier.) The exploration of infinite possibilities and the selection of one's unique profile is, in Jungian terms, a life-long process of individuation. In each person's orchestration of activities, he or she must balance a tendency to retire to the inner realms with an equally powerful inclination to engage with others and the external environment. Tilting too much in either direction results in distorted development, which in the social realm was labeled by Jung as extreme introversion or extroversion.

The process of interaction never ceases. As we have seen, some level of awareness exists whether one is awake or asleep. Full individual development requires active engagement with one's entire spectrum of consciousness, including dream states and all so-called psychic phenomena. But there are no periods when the being's local consciousness totally goes off duty. Therefore the individual defines his or her integrity in each situation. Everyone is always responsible at some level for his or her actions: pleas of nonculpability by reason of insanity, stress, emotional trauma, etc., are basically pretexts, not acceptable justifications.

If a group or society teaches its members that there are acceptable excuses for any type of abusive behavior toward others, it takes on responsibility for all such actions and is therefore culpable. Laws and regulations (such as easy bankruptcy options, corporate protections, and welfare loopholes) that permit some people to escape responsibility for their behavior encourage lack of integrity in all, and thereby weaken the moral fiber of the whole society.

Inner Gyroscope

Given the permeable nature of personal membranes, every being requires a focusing mechanism that maintains its cohesive core. What is the inertial guidance system or gyroscope that provides a being with that stability in the midst of an ever-changing life?

Hardly anything short of actual destruction of the biological envelope permanently severs the functioning of a being's consciousness from the phenomenal realm. A being can experience severe emotional losses, comatose states, physical trauma, heights of ecstasy, loneliness, public acclaim, and all forms of social slight and injury and, through it all, remain self-contained and capable of remembering who and where one is in space-time. A being can also leave its material body, travel to other dimensions, and then recover it, picking up where it left off.

Memory is a manifestation of that inner gyroscope, making it possible for beings to chart their course through the continually fluctuating state of ordinary existence. Various concepts about the dynamics of memory have been posited, but none has been proven to a disinterested observer's satisfaction. Freud's concept of repression as the process by which data is placed in memory banks and "sealed" from full consciousness is not very useful. We have not been able to validate his realms of the unconscious and subconscious because validation requires their subject matter becoming conscious; once in the field of consciousness there is no confirmable trail to prove where the matter has been.

Physical scientists hypothesize that memory is rooted in neural circuits in the brain. According to this perspective, memories are stored in varying electrical patterns, based on the "charge" of the event, in specific sections of the brain. When these sections are stimulated by similar electrical charges, some memories are released. But such relationships do not prove the memories are biologically determined. The theory is at best only descriptive of physiological relationships, unconnected to ideational content. It offers neither a rationale for discrepancies between one person's memory and externally confirmable data, nor does it account for different third-party memories of the same event.

Given such untestable theories, it is intriguing to shift the frame of reference to one that entails a different concept of memory—a multi-dimensional manifestation of conscious awareness. A useful new concept must explain not only personal memories, but also how beings contribute to the collective memory and receive from it with ease. It should also account for the existence of differences in personal perception and cryptomnesia, the process of incorporating another's experience or ideas into one's own memory. The concept of *noumena* introduced here, within the three-faceted model of reality, points in a constructive direction by equating memory with a being's idiosyncratic repository (local concentration) of material from a more general consciousness.

If personal memory is both local and linked to the noumenal whole—the holographic concept again—then it could include the phenomenon of shared knowledge. In the individual's interaction with the *noumena,* information can be received from it and contributed to it. Therefore, retrieving memories (personal or cosmic) is not unlike dreaming or psychic knowing: it involves tuning into the relevant frequencies.

The observed ability to consciously direct and enhance various channels of access to local and universal memories indicates that modern science has been studying the brain backwards. The continually remanifesting brain does not create conscious activity: it is a tool to help filter subtle signals and control the level of five-sense awareness. Given the capability of conscious beings to access all past and current events, the brain is needed to maintain order among the data from the permeable membrane that envelopes a *beingness*. Incarnated beings select the sector or wave band that they wish to attend and use their brains to define and protect themselves.

Multiple Personalities

Grounded by the gyroscope of local memory, each being has the capacity to take on different roles and to temporarily differentiate self along any behavioral spectrum. Such range of performance is applauded in actors, but in is often labeled a disease when it does not seem to be under obvious control.

Cases of what has become labeled a "multiple personality disorder" (MP) are now widely known. Productions such as *The Three Faces of Eve, Sybil*, and *The Minds of Billy Milligan* have exposed the general public to this syndrome, where one physical being serves as the "house" for a number of personalities that individually dominate overt behavior at different times.

The differences among an MP's personalities can be striking: they can be male or female, adult or child, linguistically different, varied in philosophies, and diverse in value systems. Practically any characteristic that a personality can exhibit to distinguish itself from others has shown up in these apparently unintegrated beings. Psychologists and other health care professionals have concluded that the MP syndrome results from early childhood trauma, involving emotional, physical, or sexual abuse that results in "dissociation." Dissociation is seen as a reaction separating one area of mental and behavioral activity from another, i.e., a breaking up of awareness. Perceived as a defense mechanism, it is thought that abused children use dissociation to protect certain parts of their awareness by denying the existence of other parts of their reality.

For the MP syndrome, it may be more helpful to replace the construct "dissociation" with the concept of "underdeveloped skills in integration and orchestration." Dissociation implies the result of some external force over which the individual can exercise no influence. A more appropriate concept, recognizing that conscious beings assign meaning to all external influences, would reflect the inherent impulse conscious beings have to expand awareness and integrate ever-increasing degrees of complexity. Only stifled opportunity, physical bondage, or self-maintained psychological constraints can inhibit this process. When young children who have been abused do not receive assistance in managing their own integration, the MP disorder can result.* The "cause" in this instance is not solely the abuse, where perceptions

* Some research has correlated tendencies toward schizophrenia and depression to factors that affect foetal brain development, but it is necessary to recall that correlation is not causation.

vary from one abused child to another, but also the lack of integrative skill development and/or support from others in the adult community. Although abuse is a behavior to be proscribed, with community effort to heal the abuser, the main focus of community efforts should be on assisting both individuals in the formation of self.

MPs point the way to greater understanding of the incarnation of consciousness in a human form. Research data demonstrate that aspects of the MP's physical body actually change when the personality changes. There may be variations in voice, posture, color sensitivity, left- or right-handedness, as well as changes in brain-wave patterns, immune system responses, and dermal electrical responses. Documented cases include changes in warts, scars, and rashes. A relatively simple but dramatic example is the turning on and off of allergic reactions as the personality shifts. Medical science claims a person cannot turn off allergic responses at will, but MPs do. It they can, why can't others develop the same powers?

If a temporary and partial shift in consciousness can have such visible physiological impacts, imagine the influence deliberate shifts in a well-integrated personality could have on a person's physical condition. MPs seem to have a very high self-healing capacity; they demonstrate more clearly than others the interaction of mind, energy, and matter discussed in Chapter 4. All people influence their physical health every day, but for the most part as a function of habit. Conscious manipulation of one's repertory of personality traits demonstrates the power consciousness has over its own incarnation. Acting as if something is true makes it so.

Any state experienced by an MP can be found in the mental repertory of persons who practice mind control, biofeedback, hypnosis, meditative trances, or other techniques that evoke altered states of consciousness. Therefore, the path to wholeness, or progressive integration, for *any* cosmic being—whether labeled MP or not—is the bringing into awareness all fragments of his or her consciousness and accepting them as facets of existence in this incarnation. A responsible human family or other support group responds to dysfunctional fragmentation in young beings by providing a social environment that is conducive to wholeness and integration. Mature beings model the skill of MP development, showing how all facets of behavior can be selectively used in some settings. Young beings learn in the process of socialization to use discretion in their public personality, exhibiting only behaviors appropriate to the role being played. They should also learn not to dismiss any of their repertory as unreal. Developing beings recognize that disowned parts of self absorb a detrimental share of available subtle energy, inhibiting the realization of their full potential.

Opening Self

Paradoxically, the antidote to multiple personality fragmentation is greater openness of a conscious kind (or voluntary dissociation). Individuals determine, either on an instance-by-instance basis or through established patterns, their core priorities on how much to take in from any of the physical or subtle senses. To be fully alive is to be consciously engaged with phenomenal and energetic stimuli in a manner that is consistent with a thoughtful sense of priorities. If beings deaden the capacity to feel pain (physical, emotional, or mental), they correspondingly reduce the ability to sense pleasure. The same principle applies to the polarities that characterize all emotional gradients: love/hate, attachment/detachment, fear/confidence, etc. The range of behaviors available to any being is a function of one's ability to experience both ends of each spectrum, which, as shown below, is more like a circle than two segments of a line.

NATURE OF POLARITIES

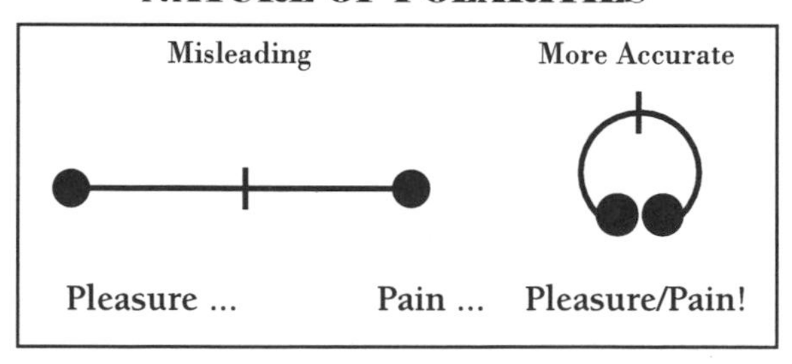

Both extremes are energetically close together. The circle model shows why it is so easy to transmute one pole of energy to another, as in flipping to sobs from laughter or vice versa.

The key to self-management is learning how to consciously ascribe meaning. A conscious being can consider a physically or emotionally painful experience to be one of the following: (1) a dream, (2) an unfelt part of oneself, (3) one's fixed core, or (4) a routine experience for an incarnated being. The greater the gap (dream, unfelt, or routine) designated between the painful event and one's immutable core self, the easier the wound is to bear. These forms of voluntary dissociation are used to advantage by people undergoing torture or other painful treatment.

A painful event first designated as a dream can later be more easily relabeled as reality (an approach used in hypnosis with UFO abductees recalling invasive operations and with veterans reliving battle wounding). To put the event on the shelf of dream enables one to function temporarily in the world of physical senses, but if one is to grow, each significant experience must be fully integrated into the whole self. The definition ascribed to an event depends on one's sense of self and assumptions about the nature of life. Conscious beings with a high degree of control over those definitions can literally shape the reality of each experience.

As discussed earlier, a local incarnation of consciousness has receptors or senses serving each of the three bodies. Dreams and daydreams, ritual behaviors (verbal or otherwise), deep meditation, and trance are all ways to expand the receptivity of the subtle senses. To the extent that one synthesizes them with the five-sense mode of awareness, mastery of the potential of an incarnation is strengthened. For example, meditation (a process for transcending the physical senses) does not deny the reality of incarnation, but enhances it.

Even though meditative concepts and techniques are most frequently associated with Eastern religions, they have been known in most cultures for millennia. Until about 200 years ago, meditation was an active component of Judaism.[10] In the Kabalistic tradition, meditation was viewed as the process by which one isolates self from uncontrolled mental stimuli and taps into generic consciousness. This act of filtering out local activity and awareness gives one conscious access to the noumenal logos. Spontaneous trances, as reported by many inventors and scientists, can provide access to the richness of the noumenal realm, but the deliberate practice of meditation can have the same result.

Deliberate inner searching is not the only way to expand the membrane of local consciousness; other techniques, such as hypnosis, are available. Hypnosis helps to bring material from the noumena or collective consciousness into awareness. It assists a person in recovering thoughts or experiences that have been temporarily pushed from the field of awareness, due either to the passage of time or their painful nature. Under hypnosis, the individual gives the hypnotist permission to probe the *noumena* through the subject's brain/mind channel of local memory.

The ease of memory recall, in any state of consciousness, seems to be related to the magnitude of the subtle energy charge attached to the idea or experience. The higher the energy load, the easier and clearer the recall, unless a countervailing, equally energized belief or memory* has gained ascendancy. When a countervailing memory exists, the dominant energy polarity must be transformed before a shift to the out-of-phase memory is possible. A good hypnotist or shaman knows how to do this, but any conscious being can develop self-guiding techniques for transmutation of energy polarities from one half of the circle (see previous graphic) to the other. These techniques involve controlled bursts of energy, deliberate redirection of flow, or personal commands to oneself.

When access to an area of memory or learning is blocked by an opposing polarity, humans usually feel frustrated. Sometimes, however, they feel threatened by the unknown behind the block. Even under hypnosis, a subject feeling threatened in this manner will resist probing questions. When such energy blockages are converted into open channels, more and more related data from the restricted area flows through. It is similar to a situation when despair is converted to hope and all the reasons for hope become accessible to the awakened mind.

---

* Such countervailing memories can apparently be artificially insinuated by other beings (human or ET) through the power of suggestion when the subject is sufficiently intimidated.

External Confirmation

The opening of self gives a being access to more noumenal and energeial information, but that does not mean it is all valid. Truth cannot be easily distinguished from fiction through use of the ordinary awareness; the same is true for the subtle senses. The input from any one or combination of senses should be confirmed through other channels before meaning is ascribed. Since the noumenal borders between personal experience and that of others are so porous, care is needed in interpreting all such communications.

To take one area, in hypnosis, to illustrate, care must be taken to insure only personal memories are being tapped by the hypnotist. Memories recovered in hypnosis will be misleading if they are influenced by the hypnotist's suggestion or by information bleeding through from the impersonal *noumena*. In the former one may be led to create an event that did not happen and, in the latter, the experience of another may be read as one's own. Externally verifiable data should be sought to at least partially confirm the hypnotically reported experience. For example, in the case of the UFO abductee, obtaining field evidence (physical traces, scars or other body markings) or confirmation of part of the experience by third parties* is crucial for credibility and subsequent research.

The same caution should be applied to all subtle-sense ways of knowing, for having a channel to knowledge does not guarantee the validity or relevance of the information received. When beings have access to anything in the *noumena*, they must be as careful using it as using gossip conveyed over the garden fence. Information in the noumenal realm can be just as deceiving or misleading as various aspects of the phenomenal realm.

The following incident exemplifies the way an apparently real memory or bit of knowledge can be based on something unreal. As a tool in therapy with a patient, psychologist Milton Erickson[11] once created a fictional man. After regressing the patient back to childhood, he suggested the memory of

---

* People who consciously recall the same events that others have repressed can be used for cross-validation of hypnotically retrieved material.

a person she might have known and created a character for her. In a subsequent therapy session he regressed her to childhood, using the fictional being as a participant in the resolution of the disorder. Upon waking in her adult consciousness, the patient could not be talked out of her belief in the reality of that imaginal person.*

As more and more beings become aware of the potential for multiple paths to knowledge, there is a need to encourage self-discipline in interpretation. Individuals wishing to be responsible can join in establishing a process of intersubjective validation. To gain a higher degree of confidence in one's subtle knowing, an individual can arrange to test interpretations by setting up predictive situations where a third party has advance access to the expected outcomes. In one example, the individual commits to writing a precognitive knowing, or informs another of it, and asks the third party to compare the subtle sense message with the subsequent event. This way, one avoids misleading oneself about the accuracy of the precognitive perception. There is a strong temptation to retroactively shape one's memory to fit the circumstances that come to pass. By asking others to verify or duplicate one's own experience, it is possible to gain a degree of correlation, if not absolute validation, of the communication.

Since the subtle senses are largely invisible to third-party, five-sense observation, and often subject to ambiguity, it is important to realize that frequently so-called objective validation is not possible. Given the now broadly understood observer effect, where the manner of observation shapes the event, even *matenergy* events cannot be objectively verified. Therefore, intersubjective validation is the only recourse we have, i.e., at least two beings separately test their perceptions and compare the extent to which they are in agreement. This approach makes it possible to be scientific about things not normally considered to be in the realm of science, facilitating the expansion of scientific knowledge to include much that has been considered paranormal. An excellent way to test this approach is through subtle energy healing that involves observable physical results.

---

* As a graduate student, I witnessed a similar experiment—performed by our clinical psychology professor with his children—that taught me this lesson well.

Self-Healing

Each being has an inherent self-healing capacity that works at all levels, physical, mental, and emotional. In other words, a living system has an innate impulse to wholeness: when one of its elements is weakened by an external attack or by inadequate inner nourishment, it reacts automatically to protect its integrity. Only unabated aggressive invasion or perpetual lack of inner stability will be able to overwhelm the natural defense systems. Unfortunately, many cultural patterns (aggressive legal and business norms, toxic pollutants, and incessant images of violence) undermine these natural cycles and rhythms and result in dis-ease.

Treatments that run counter to the system's natural principles or induce deleterious side effects may serve as short-term palliatives, but they will not heal the being. Only interventions that arise from and support the natural processes will eliminate the causes of symptoms. This is true whether the problem is a headache, sleeplessness, cancer, heart disease, AIDS, neurosis, psychosis, moral turpitude, inner city riots, ethnic warfare, or spiritual malaise. The Principle of Correspondence insures that illness at any level will be manifested in all three facets or bodies. Conversely, the commencement of an appropriate cure in any one system will have positive ramifications for the whole organism.

At the simplest level, the physical body's response to microbial invasion is to defend its health. Some of those defenses appear as symptoms that modern drugs are designed to "cure." But each defense has its purpose. The fever humans hasten to suppress is stimulating the growth of white blood cells to inhibit the growth of bacteria. Nausea in pregnant women protects the embryo from getting trace poisons through the mother's diet. A headache is a signal of too much stress.[12] If such defenses are not properly allowed to run their course, other parts of the organism may suffer. Anxiety attacks treated with tranquilizers will recur as long as perceptions of dangers continue. When a person—failing to listen to the body and appreciate the Principle of Cause and Effect—acts inappropriately or without full awareness, he or she impedes the self-healing process.*

Modern medicine portrays addictions (to alcohol, drugs, nicotine, etc.) as only physiologically based, rooted in cellular and organic dependencies that can be satisfied by infusions of a particular substance, or its artificial substitute. But as *consciousness is the predominant plane*, physical dependencies reflect patterns either initiated or accepted by the individual being. An aspect of evolving *beingness* is the gaining of an understanding of the power local consciousness can bring to bear on the dependencies of the physical body.

Thought can create and modify basic configurations of matter, enabling a strong being to determine its degree of dependency on any particular substance. Smoking, drug use, and other addictive behaviors can be broken by a change in thought: *anyone who desires can change the underlying thought pattern of addiction with the appropriate support of other beings.*

Having been deluded into thinking that health and well-being are attainable through expensive prescriptions and over-the-counter drugs, humans have lost awareness of their significant capacity for self-healing. They no longer remember that cycles of activity must be balanced with rest and recuperation. They do not perceive symptoms as messages bodies send minds, and vice versa. With the channels of communication between realms silenced, humans have experienced fragmentation among their constituent parts.

A profound understanding of the nature of *beingness* or consciousness incarnate is recognition of its inner capacity for self-healing, self-learning, self-healing, and self-maintenance. Humans who are inattentive to being an organic part of a conscious universe suffer the consequences, and diminish the potential performance made possible by their cosmic creative powers.

---

* The pharmaceutical industry in the U.S., along with narrowly focused physicians, has used the power of the media advertising to promulgate the notion that the body is a machine which can be kept in tune by the ingestion and application of chemicals. Every drip of the nose must be artificially dried up. Pains that signal something is wrong and needs conscious attention are anesthetized. Blocked bowels that cry out for nutritious food are treated with lubricants. Indigestion caused by stress is masked with chemical foam. Skin cells that need water are salved by external lotions.

NOTES

1. Scholastico, Ron. *Doorway to the Soul* (Scribner: New York, 1995).
2. Grof, Stanislav. *The Adventure of Self-Discovery* (State University of New York Press: New York, 1988) and *The Holotropic Mind* (Harper San Francisco: San Francisco, 1992).
3. Harner, Michael. *The Way of the Shaman* (Harper & Row: San Francisco, 1980).
4. Monroe, Robert A. *Far Journeys* (Doubleday: New York, 1985).
5. Redfield, James. *The Celestine Prophecy* (Warner Books: New York, 1993).
6. Montgomery, Ruth. *Strangers Among Us* (Ballantine Books: New York, 1982).
7. Gerber, Richard. *Vibrational Medicine: New Choices for Healing Ourselves* (Bear & Co.: Santa Fe, New Mexico, 1988); Clow, Barbara Hand. *Liquid Light of Sex* (Bear & Co.: Santa Fe, New Mexico, 1991).
8. Clynes, Manfred. *Sentics: The Touch of Emotion* (Anchor Press/Doubleday: Garden City, New York, 1978).
9. Cooper, Vicki. "Interview of William Cone." *UFO Journal* 7, No. 2 (1993).
10. Kaplan, Aryeh. *Meditation and the Bible* (Samuel Weiser, Inc.: York Beach, Maine, 1992).
11. Erickson, Milton. *The February Man: Evolving Consciousness and Identity in Hynotherapy* (Brunner/Mazel: New York, 1989).
12. Williams, George C., and Randolph M. Nesse. "The Dawn of Darwinian Medicine." *Quarterly Review of Biology*, December 1991.

PART III

Humans in Co-Creation

PART III PORTRAYS METASCIENCE as a catalyst for a new planetary renaissance and challenges humans to join the cosmic fraternity of all beings. It explores humanity's role in the process of co-creation in the universe, uncovering the greater potential of humans made possible by conscious awareness of their potential in the cosmic community. Growth of personal capacities and skill in conscious living is seen as resulting from the developmental dynamic between self and other, including communication with other species and the living planet itself.

Chapter 7

Potential: Cosmic Humans

Conventional wisdom has a limited view of humans and their history—a recent and unique life form added to planet Earth by a haphazard process of evolution. Frontier metaphysicians and historians have a widely differing perspective—descendants of a long chain of civilizations that have included contributions from societies far in advance of twentieth-century industrial society. In this schema, humans are the provincial actors in a cosmic cast of beings arising out of universal consciousness. To know ourselves requires re-discovery of this consciousness of which we are part. Before humans can aspire to galactic stardom, they must grasp their own identity and its origins.

◆ ◆ ◆

GENERALLY, even the most educated hold vague ideas about the origins of human culture. Most scholars assume a few legendary men gave birth to advanced civilization relatively overnight, attributing to these untutored individuals subtle ideas of law, justice, responsibility, freedom, honesty, beauty, goodness, and service. Since forms of government and institutions of cooperation, altruism, and defense seem to have blossomed in a relatively brief period, scholars have unquestioningly accepted that such achievements in consciousness could spring full-blown without antecedents from Stone Age and Iron Age tribes. To remain psychologically comfortable, these scholars appear to have suspended their critical faculties with regard to such assumptions, somewhat like children with a vested interest in perpetuating the myth of Santa Claus.

Blinders to Knowing

Since much information from records less than 10,000 years old is not admitted into "official history" (see Table XI), it is not surprising that

Table XI

VIEWS OF EARTHLING'S HISTORY

Possibly	Plausible	Esoteric	Scholarly	Conventional
More Far Out				
Himalayas	Far Back			
65 + million				
Colombia				
63 million				
Nascans of Ica				
50 million				
Peru				
5 million		Even Further		
	Abzu - 300,000			
	(Adapa/Adamu)			
	Apsu-200,000			
	(Chava/Eve)		More Ancient	
		Eden - 70,000		
		(Gilgamesh)		Ancient
		Zambesi - 50,000		Civilizations
			Sumer -12,000	
			Hindu - 11,500	
			Egypt - 11,000	
			Babylonia - 11,000	
			China - 10,000	
			Atlantis - 10,000	
			Mayan/Olmec - 8,500	
			Akkad - 3,000	
				Egypt - 3,000
				Crete - 2,500
				Assyria - 1,912
				Babylonia -1,900
				Phoenicia - 1,500
				Greece - 1,100
				Rome - 753
				Persia - 559

evidence from earlier periods is ignored. That anything more than a few thousand years old is deemed "prehistory" demonstrates the limited nature of the academic perspective. This narrow outlook on history also constrains study of present human capacities. Table XI offers a much more expansive view of our possible history. The Western method of scientific inquiry, including history, which revitalized human thought during the last Renaissance, has now boxed itself in: powerful research and academic communities ignore most prehistoric and much historic human experience. In his book *Exploring Inner and Outer Space*,[1] Brian O'Leary clearly demonstrates how the theoretical and experimental box of established (fundamentalist) science is inadequate to deal with a picture of reality that encompasses all the evidence of human history and potential.

Since the seventeenth century of Francis Bacon and Isaac Newton, mechanistic science has become the dominant perspective of human thought. With its inductive method of logic, observation, experimentation, and confirmation, the so-called modern scientific method has become the criterion by which new ideas and traditional wisdom are measured before they are classified as valid. The more subtle aspects of human experience have been pushed off the superhighways to roads less traveled. It is now up to the current generation of "new scientists" to integrate empiricism with experiences like intuition and mind/body healing, thereby opening new pathways for understanding and enhancing human experience. Researchers need a conceptual framework for science that goes beyond the current materialistic boundaries they are forced into.

Unfortunately, established science is not the only institution whose limited approach to learning restrains human progress. Religious fundamentalism is another; it asserts that an anthropomorphic god manages the universe and human affairs in a personal manner that is miraculous and not subject to preestablished principles. For true believers the answer to any human question is, "because God made it that way," or "Allah did things this way for his own reasons which we are incapable of understanding." Such attitudes in most religions are even less conducive to new learning and intellectual growth than those of institutional science.

As a polarity to fundamentalism in science and religion, there is an alternative science in every generation. Called "new scientists," its proponents work outside the accepted boundaries, challenging society to support the flowering of broader perspectives. Although the International Association for New Science was not founded until 1990, "new science" was used in the 1920s and 1930s to denote those who were ahead of their time in the fields of consciousness and mind/body interaction. And that movement was in turn built on the theosophical and "new-thought" trends of the late nineteenth century. Aspiring scientists should see themselves as part of a river of discovery and elaboration. The practice of standing on the shoulders of one's predecessors operates in every field.

For example, Cleve Backster's recent work involving intercellular communication between species, discussed in Chapter 2, has broad shoulders from the past. J. Chandra Bose of India was knighted by Britain in 1917 for his invention of the crescograph: it demonstrated that plants have nervous systems and respond to emotional stimuli from humans. Before the turn of the century, several people were laying the groundwork for Einstein and twentieth-century physics. In 1908, Minkowsky conceptually united local space and local time into an absolute four-dimensional space-time, but he too was building on earlier work, that of Edward Morley, George Fitzgerald, and Hendrik Lorentz, all of whom had been working with the concepts of ether, contraction, and "local time." Some roots of Einstein's theory of relativity can be seen in Jainism and Buddhism from the period 600 B.C.E. to 200 C.E. The Jain story of the six blind men feeling different parts of an elephant reminds one that perception depends on where one stands.

How can scientific progression be encouraged and strengthened? Are we considering all the sources of knowledge available to us? What is hidden in the past that could help light the way to a more fully lived future? Is telling evidence of the Solarian Legacy being ignored because of self-imposed blinders? Does avoiding these questions limit understanding of what it means to be a human being? We must start by understanding the triadic nature of humans in the larger field of consciousness.

Cosmic Beings

As we have seen, scientism—a form of pseudo-science more motivated by dogma than the search for truth—and religionism—again, more emphasis on dogma than on truth—have unwittingly conspired to increase confusion about the essential elements of the universe and human beings in particular. Simplistic assertions about humans that are only dogma, unable to stand the test of rational examination, are frequently cloaked in arcana to encourage the blind faith that is required to believe them. Often fuzzy terms find their way into popular usage without a general agreement on their meanings, e.g., *spiritual realm*, *divine level*, *astral body*, *intuition*, *psyche*, *subconscious*, *unconscious*, *preconscious*, *soul*, *core self*, *energy body*, *physical organism*, *brain/mind*, *psychological illness*, and so on. Do they accurately distinguish different levels of being and states of awareness? Can one collect data about these categories and be sure the same data is not subsumed under another label by someone else? The terms are so lacking in clear distinctions that it is no wonder communication is so difficult about the subtle, and frequently labeled "anomalous," aspects of human experience.

Ultimately, the cosmic reality of importance to humans is inherently simple and intellectually easy to grasp. Following the principle of Occam's razor (that the simplest explanation is the most plausible), it is possible to use ordinary words and phrases, with meanings about which there is general agreement, or to provide clear definitions. Taking this approach, the experience of a human being can be reduced to three areas: *physical, energetic,* and *mental*, not unlike the body-spirit-mind triad used by some. They reflect the three realms (*phenomena*, *energeia*, and *noumena*) and their corresponding bodies discussed in Chapter 4. While relatively distinct, but still interdependent, these three facets provide explanations for all human experience.

- **Physical**: Either waves and particles or energy and mass, it includes both inorganic and organic forms *(Phenomena)*
- **Energetic:** The realm of feeling/emotion and subtle force, it involves states of excitation beyond material parameters *(Energeia)*
- **Mental**: The realm of knowledge and abstract ideas, it exists beyond ordinary awareness *(Noumena)*.

Any aspect of human experience can be seen as a subset of one of these categories, a combination of two or more, or the dynamic interaction among the three. All human activities involve this threesome, analogous to the triadic structure of the three-quark building blocks at the proton and neutron level discussed in Chapter 2. (This isomorphism may explain why the concept of "trinity" is so common to ancient, but advanced cosmologies.)

Although the three facets are functionally interdependent, mind or local consciousness is *primo inter pares*, the charioteer that directs the horses that pull the chariot. Mental patterns are the only forms susceptible to direct, conscious manipulation by humans or other incarnate beings. Inherent constraints or natural laws set fewer limits in this realm where the degrees of freedom are much wider than for either energetic or physical activity. Mind has great scope for creating new patterns, but if it wants to change the phenomenal realm it must work through the laws governing intermediate forces. The way new images in modern sports have expanded the boundaries of physical behavior exemplifies this principle.[2] What started with "inner tennis"—the visualization of perfect form serving to focus the athlete's energy—has now led to performances that each year set new heights of achievement in most sports. The ancient practice of Tantric Yoga is another example, wherein mental images of multilevel sexual activity help expand energetic and physical limits. At another level, psychokinesis, directed thoughts, can move subtle energy and thus affect matter at a distance. In the most powerful case, conscious beings can harness subtle forces to cause the materialization of new objects.

The concepts of physical and mental bodies for a life form are currently more easily understood than is the energetic body: the popular connotations of the first two closely parallel their scientific meanings. The problem with the term "energy body" is that the common use of the word "energy" refers to the obverse aspect of matter, the wave instead of the particle. In organic entities, the energetic body is not the same as electromagnetism and gravity. The energy body is composed of the *subtle energy* forces in the conscious being that mediate ideas into form and vice versa.

Over the ages this subtle energy has been given various labels: prana (Hindu), holy spirit (Christian), chi (Chinese), and ki (Japanese). In the history of Western science it has been conceived of as magnetism (Mesmer), orgone energy (Wilheim Reich), or an electric current (Becker). Now frontier physicists postulate psi forces, tachyons, Bell correlations, or collapsing wave functions. Physicist David Peat sees it as a force in the process of forming (clinging to form) in the material realm. Regardless of the label, recognition of reality demands it be analyzed and incorporated into thinking about the nature of human experience.

Michael Riversong[3] has theorized that such waves (whether they are known by the above terms or as zero-point, scalar, or organic energy) propagate exactly like sound waves. He points out that much of the Sun's energy is in the form of sound. However intriguing the thought that all natural processes produce low-volume sound waves at all frequencies, that such activity is subtle energy does not necessarily follow. (Refer again to the postulates of other spectra in Chapter 4.) In the framework presented in this book, the energies referred to by these labels fall into the energeial realm, and are not the same as sound in the phenomenal realm. The subtle energies, like *matenergy*, do not act on their own: they require direction from the charioteer. Contrary to popular belief, feeling and emotions follow the orders of thought.

Although scientists have learned a lot about how nerve cells called "neurons" send signals and how the signals in chemical form, called "neurotransmitters," pass from one cell to another, they still cannot explain how such chemical signals give rise to thoughts and feelings.[4] Can love and hate, ecstasy and despair, really be the result of random nerve cells passing molecules back and forth?

"Frontier science," as articulated at the center known by that name and founded by Beverly Rubik at Temple University,[5] is now recognizing the possibility of "an acausal mind-matter interrelationship that is fundamental."

In other words, mind and matter are believed to coexist interdependently. In the three-faceted model, they are bonded by the field of subtle energy or *energeia*. Within this framework one can see the possibility of reciprocity between consciousness and various behaviors. After the link has been identified in theory, it is possible to test the directionality of these influences. The evidence of human experience indicates that the power to override matter resides in consciousness, notwithstanding that local mind can be influenced by matter in their reciprocal communications. The fundamental question is, "from where does that mental power come?"

Local Mind

Cosmic consciousness permeates a gradient from local to universal. A local mind is the part of that consciousness incarnated in a particular being. Unlike cosmic consciousness—the field of origin—consciousness incarnate is focused and identifiable, yet undivided and indivisible from the whole. Local consciousness or mind fuses with the force through which it acts (*energeia*) and the medium on which it acts *(matenergy)*.

Consciousness, when manifesting its essence in material form, incorporates multiple senses (see Chapter 4) to enable itself to monitor its own participation in a space-time existence. The use of physical and subtle senses permits the consciousness, which stands outside space-time, to experience itself incarnate. Some conscious beings can apparently determine, perhaps in advance of birth, not to lose awareness of these connections with their cosmic origins. They continue to see themselves as transcendent beings throughout an incarnation. For others who lose this sense, the "membrane" between local consciousness and universal consciousness is always permeable to some degree, and every being can pierce it by accident or design. Since the universal is perceived through the lens of the physical or subtle senses, the very act of perception proves the indivisibility of the three realms.

Consciousness seems to follow the same laws as energy, for example, the law of conservation ordaining that energy cannot be created or destroyed. As an ideational form of energy, local consciousness can assume different states. Its expansiveness is inversely proportional to the inhibitions or prejudices that would limit flow, as in Ohm's law of electricity in which the flow of current is inversely proportional to resistance. The isomorphic nature of the mental and physical realms shows itself in another manner: the way one focuses on an electron determines whether it will be a wave or particle, and the way one focuses on an idea shapes its polarity, i.e., its positive or negative charge.[6]

For millennia, people—depending on their world view—took free-floating mental input as either natural messages or special visions from the gods. The Egyptians and the Hebrews considered Moses' dreams to be divine messages. Eskimos and Native Americans, as do most traditional people, see their dreams and unbidden visions as communications from Holy Spirits. Although visions or ideas coming at night may appear to be special messages, they sometimes, like idle reveries, are the source for solutions to a problem, or new insights or discoveries. The sea of consciousness is a source of valuable knowledge and wisdom when accurately understood.

All conscious beings swim around in this mental sea of words, images, and visions. They use it to create and express new concepts, interpret data from their multileveled environment, and engage in nuanced communications. The unintended results of these activities may be contributions to the shaping of cosmic-level reality. It is not unreasonable to infer something of the power of general consciousness from the way the beings we know best—ourselves—experience it.

Unified Source

In this century, as part of the pervasive sweep of the dualistic, materialistic approach to science, two people recast Western thinking about the nature of consciousness. Sigmund Freud and Carl Jung took largely uncharted territory and applied their particular constructs as templates over the human mind and consciousness.

Freud started with the assumption that two basically separate components—the conscious and the unconscious—resided in the individual. Jung was more inclusive in his approach, but like Freud he essentially accepted the largely biological yet dualistic model of the human being. Both projected various entities within the human being—id, personal unconsciousness, ego, self, superego, and collective unconscious—and ascribed to them largely autonomous powers.

Freud asserted that the fundamentally independent personal unconscious had desires so unacceptable to human consciousness that it disguised them through hidden meanings and symbols in dreams, "Freudian slips," or compulsive behaviors. He considered interpretations by specially trained analysts were necessary in order to decipher their meanings.*

Jung's model, with its concept of the collective unconscious, expanded the range of interpretations for both the source and purposes of such mental material. In recognizing that personal reality is much larger than that which is perceived with five physical senses, he opened the door to exploration of group consciousness, or the *noumena*. His early focus on the human dream world as the primary route into the collective unconscious, however, deemphasized the many paths of deliberate access, including meditation, remote viewing, trance, day dreaming, and use of hallucinogens or ritual movement.

Jung did use his vast intellectual capacity, focused on the role of symbols[7] in consciousness, to show how dreams relate to other spontaneous channels for knowing, like flashes of precognition and clairvoyance. He believed such involuntary acts provided fragmentary access to an individual's reservoir of data that was subliminally received from the environment or inherited biologically. But they were more "like instincts" than the conscious charioteer. He considered consciousness as a "recent acquisition of nature...still in an experimental state," and viewed civilization as necessary to establish a high level of continuity and avoid fragmentation of mind. For him dissociation was a problem related to "primitive" societies that would be overcome by civilization.

When it became apparent that material from the seeming fringes of consciousness frequently provided wise counsel or innovative insights to people, Jung posited the idea that dreams were from an independent unconscious that was somehow an autonomous "healer." Reverting to tradition, he

---

* His idea of a realm beyond our control has become so rooted in modern psychology that now people escape personal responsibility for murder and other serious crimes by claiming they were unconsciously (insanely) driven.

attributed to this dream world an almost divine purposefulness, saying, "God speaks chiefly through dreams and visions." In his view,* this realm of unconsciousness had a will of its own and could guide beings in a more balanced way through waking life, if only they heeded its message.

The psychoanalytic perspective of both Freud and Jung, treating all maladaptive expressions of consciousness as a medical/psychiatric matter, has unfortunately limited the development of consciousness research. The use of terms such as "treatment," "emotional disturbance," and "neurotic phenomena," has caused other professions to consider these aspects of self as a category best left to psychiatrists, thereby losing the benefits of interdisciplinary synthesis. Individuals can access the common reservoir of consciousness directly—without resorting to some intermediary healer or specialized concepts. Understanding can come through rational review of dream content or similar images gained through meditation and other channels.

In a universe of seamless consciousness there are no separate parts in incarnated beings that heal or guide other parts; there are only aspects of the whole that may receive less attention from the individual's focused awareness. The dreams of any being are not divine and do not come from another realm. They are both part of and reflect the current level of conscious development of the individual being. Helpers (for humans, that includes psychologists, psychiatrists, counselors, etc.) can only assist the individual in the process of self-integration, for the ultimate responsibility remains with the being.

Many who have not been persuaded by the psychoanalytic school of thought and its modern variants have turned to chemistry and biology to uncover the bounds of consciousness, which they frequently equate with ordinary energy. Their materialistic approach assumes that consciousness derives from a being's physical organs. They imagine the brain cells consolidate memories of the day's events, getting rid of some of the clutter, and

---

* Jung's profound education in history and culture enriched his interpretation and definition of dream content. Unfortunately such erudition has resulted in his devotees' creation of an international Jungian industry (books, workshops, academic courses, psychological techniques, etc.) that obscures the straightforward nature of the role of dreams in consciousness. As we have seen, dreams are only one of many channels to the *noumena*, and should be seen as other sources of data, as any sensory input, for rational interpretation.

then recharge themselves. Disparate cells are assumed to independently discern an order to events, make value judgments on what is relevant, and discard the rest. This set of assumptions is counter to the continuous and orderly nature of human experience, and likely counter to the experience of sentient beings elsewhere in the universe.

Insights from consciousness research reveal that the universal field of consciousness, not unlike other forces, can operate on various frequencies. It is pervasive throughout, and reaches beyond, space-time, continually evolving from experience. While its information can be clear in both pattern and intensity, incarnate beings can also pick up garbled bits and pieces (in dreams, meditation, etc.) analogous to the background noise or static on a poorly tuned radio. When this happens, one must make a focused effort to read order into the informational fragments (like a multilevel set of jigsaw puzzles) and to relate them to the appropriate context. In other words, a deliberately conscious observer is necessary to make sense of all material from the noumenal field of thought forms.

Human Incarnation

Incarnation is the term applied by many to explain the nature of human existence. Its use implies an exogenous origin for an essential part of ourselves that manifests itself in this physical realm. It assumes incarnation is necessary for self-awareness, i.e., a separate self needs boundaries with built-in senses to perceive that which is not it, in order to recognize itself. Many believe a more or less permanent individual consciousness—one that may supraconsciously perceive many incarnations—operates outside of our space-time. A few speculate different incarnations exist simultaneously in parallel universes in the same pool of consciousness, but such considerations are outside of the realm of this book. We can deal rationally only with the nature of consciousness of which we are currently aware.

A useful metaphor for incarnation* is a large undifferentiated plastic sheet, representing the field of consciousness, where small concentrations of force push the sheet out of its plane at various points without puncturing

---

* Introduced in informal discussion by philosopher and mathematician Phillip Metzler.

it. The resulting three-dimensional protrusions are incarnated selves. They exist as long as the unseen force is applied; but when retracted, the plastic sheet resumes its taut expanse with only slightly visible (ghostly) impressions of the former protrusions. Another helpful analogy is a crystal wine glass, formed from amorphous grains of sand to serve until melted down to be recast as a figurine. It captures three hypothesized aspects of incarnation: origins common to other beings, a transitory unique existence, and a return to the source for another incarnation. These two metaphors illustrate a process, but they leave out the factor of intent behind incarnation. What is the force that produces the plastic sheet, and who conceives of the wine glass or figurine?

Descartes said, "I think, therefore I am," thereby at least proving to himself, and those who agree with him, the existence of a thinker. So if we think of ourselves as incarnated rather than created, at least to ourselves, we are both part of the something else that thrust us here and the manifestation of that thrust. By choosing to believe it, we act in a way to fulfill the prophecy. But if we are trying to learn from experience and to improve, we will test our hypotheses by looking outside our contemporaneous and personal realities. What the potential being intends before it almost punches through the sheet is material. But the question is, can that intent be divined from what the human incarnation chooses to do in the Solarian existence? If the Principle of Correspondence is valid, one's life here must give some indication of the purposefulness of its origin. If one's creative expressions in this life are indications of one's "divinity," then we do know something of the "divine."

Incarnation implies the manifestation of consciousness, or mind, in matter, including the brain. As the key, but not exclusive, locus of mind/matter interaction, the human brain has neurons and neurotransmitters numbering in the billions. The materialist, believing that this combination of nerve cells creates the mind, would maintain that if we put enough neurons on a microchip to simulate the cerebral cortex, we would get a computer analog to the human brain that could produce the same thoughts. The materialists end up in a blind alley, unless they recognize that the mind that fabricates the chip always precedes the product.

Physical scientists, like Robert Ornstein and Richard F. Thompson, writing in *The Amazing Brain*,[8] make a special point of the brain's physical characteristics and assumed role:

> *It is about the size of a grapefruit. It weighs about as much as a head of cabbage. It is the one organ we cannot transplant and be ourselves. The brain regulates all bodily functions; it controls our most primitive behavior—eating, sleeping, keeping warm; it is responsible for our most sophisticated activities—the creation of civilization, of music, art, science, and language. Our hopes, thoughts, emotions, and personality are all lodged—somewhere—inside there.*

The materialist's *speculative hypothesis* that consciousness arose by the chance clustering of brain cells remains just that. But the opposite hypothesis is not easily proven either. The mentalist (mind-over-matter) point of view asserts it is the mind that orders the neurons in the first place. In response, the materialist asks why chemicals can affect the moods and other functions of the mind if it is independent of the material brain? The mentalist responds that the chemicals are only cluttering up the physical channels through which the mind must operate in the five-sense world. Materialists and mentalists, from two different perspectives, see the same reality: the interdependency of the mind and its material counterpart—the brain. Incarnation is thus not a simple matter of a freewheeling mind lodging itself in just any physical host; there is obviously a complementary and interactive relationship.

Humans clearly exist at some stage between having total control of mind over matter and being totally subject to matter. We can exercise the power of psychokinesis in certain ways. We possess a consciousness that can engage in visualization and positive thinking to enhance our health. The thoughts and feelings of the mind ordering up neuropeptides to handle pain is a powerful example of their ability to manipulate *matenergy*. Yet humans can neither directly convert energy from outside ourselves to sustain our own lives, nor convert inorganic matter to organic matter in a process of direct manifestation. We have to depend on plants (or animals that in turn survive on plants) to transform inorganic elements (liquids, gases, and solids) to usable fats, proteins, sugars, and carbohydrates. We are dependent on other beings to conceptualize and conceive us. It therefore seems reasonable that the creator of our limitations has more power over the interaction of matter and consciousness than we do.

The question remains whether one day humans will be able to independently manifest themselves and inanimate objects through the assimilation of one or more cosmic energy sources. Will the strengthening of human subtle senses make such manifestations feasible? Will an expansion of conscious awareness incorporate these powers? We cannot answer these questions yet.

Mind's Power

The public wave of interest in the healing power of the mind, both symbolized and reinforced by Bill Moyers' 1993 PBS series on the topic, is bringing to light widespread evidence in the form of personal stories of self-healing powers. Popular books are now available from people who have cured themselves of cancer and other diseases after medical professionals have given up hope. Many doctors dismiss these unscientifically explained healings by calling them spontaneous remissions or anomalies, but growing numbers of professionals, as well as lay people, are attributing such phenomena to conscious intent. For two clear treatises in this area, see Elliott Dacher's patient-oriented books that reveal a physician's experience and insights.[9] The mind/body connection in healing is now beginning to get mainstream attention.* The U.S. Congress and National Institutes of Health are beginning to take it seriously by funding the alternative medicine research projects mentioned earlier.

Plain talk about dreams is another way to illustrate power of the mind over physical nature. Diametrically opposed positions such as those of Stephen Laberge (expert on lucid dreaming), who demonstrates that humans can be aware of and participate in their dreams,[10] and Francis Crick,

---

* The author served as a panelist at the WorldMed '96 Congress on complementary therapies in medicine chaired by former U.S. Surgeon General C. Everett Koop in Washington, D.C., in May 1996.

who argues that dreams are but excess data dumped from the physical brain, leave the layperson confused about the role of consciousness. The implication of Laberge's hypothesis is that we can deliberately enter into a *separate* realm and manipulate it, while the biological argument is that both conscious thoughts and unconscious dreams are created by the *same* cerebral neurons. What both theories unwittingly demonstrate is that consciousness incarnate is indivisible from, yet capable of acting on, different points on the mind-energy-matter spectrum. Practicing biofeedback to shift brain waves and visualizing to create cellular changes are two other methodologies using the subtle senses to exercise the greater power of the mind over emotional energy and matter.

The examples of the power of the human mind over matter reviewed in this book indicate human consciousness is apparently more powerful than the other Earth species'—although we cannot be sure—but there is no clear sense of how human consciousness fits into the hierarchy of cosmic consciousness. There is no overall consensus as to what the hierarchy is. To recognize that there are various levels of development among conscious beings provides no clue as to where humans are in the structure or how they got there. It does not answer the question of whether humans progressed unaided to the current level of self-awareness on this planet, immigrated here, or were catapulted from a less aware state to a more fully conscious one by the intervention of more advanced beings. Two frontier scientists have an intriguing perspective on how the current human mental capacity came about.

Cabrera (discussed in Chapter 3) believes colonizing beings implanted cognitive codes in Earthlings through insertion of sets of nucleic acids and proteins into the cerebral cortex, and perhaps then reinforced the coding through electromagnetic effects. He postulates that such biochemical interventions or manipulations of the genetic code enhanced the capability of primitive Earthlings to process knowledge. Sitchin, working independently of Cabrera with data from the other side of the world—in E-Din in what is now Iraq—also describes evidence of ET gene splicing that moved prehistoric Earthlings to a level of greater mental capability.

Although gene splicing by more advanced beings would help explain Earth's confusing archaeological record (such as the coexistence of *Homo erectus* with *Homo sapiens* when the theory of evolution says the latter came from the former), it does not appear, on the surface, to be compatible with the idea of a pre- and postlife existence of an individual consciousness. Were there advanced beings-in-consciousness without access to advanced bodies in which to incarnate before ETs implanted genes that set our ancestors apart from the chimps? The answer awaits evidence showing that if we help build a new type of body, a recognizably different conscious being will come. (The mastery of cloning techniques will also provide the opportunity, if we dare take it, to test the concept that individualized local minds are independent of their genetic heritage. Will a consciousness truly different from the cell donor's inhabit the physical body derived from the donor's gene patterns?)

Humans and chimpanzees share more equivalent DNA sequences in common than either does with gorillas, orangutans, or the great apes.[11] If, from a physiological point of view, humans and chimps are so closely related, what accounts for the intellectual and psychological differences? Do slight genetic variations account for differences in brain size, digital flexibility, image manipulation skills, hair pattern, and skin pigmentation? If they do, which takes precedence, artificially arranged gene patterns or preexisting conscious desires? Should we try gene splicing with chimpanzees, in an attempt to have them approach our level of consciousness, to test the theory of manipulation? Regardless of how humans reached their current level, it is now clear that a consciousness more powerful than we have imagined is the birthright of every human. Their Solarian Legacy has poised humans for a much greater role than they have been living.

An elementary understanding of the dynamics of incarnation and its powers still does not make it clear why humans are here. The Hindus and Western psychics, such as Edgar Cayce, have portrayed the incarnation of a particular soul in a particular body as the result of Karma, or the Principle of Cause and Effect, at play beyond local space-time. In this theory the being newly incarnated on the planet is faced with experiential challenges to overcome and opportunities for growth. Successive rebirths are required for the individual soul to balance its positive and negative reactions to the universe's imperative to grow and contribute to cosmic progress. Past life memories are seen by some as confirmation of reincarnation; but given human ability to access the *noumena*, including shared memories, recollection of events and people from earlier times does not necessarily prove reincarnation.

In a corollary to the idea of a Karmic purpose, most religious cosmologies include a belief that human beings are on Earth to work out something before moving on to the next stage. The Judeo tradition indicates Yahweh expects humans to follow "the light" to be ready for an ultimate day of judgment. Christians must ritually accept spiritual salvation from a "divine" being to have access to the realm of the gods. Buddhists see each incarnation as another challenge to achieve nirvana. The Sumerians believed they had been left on Earth by the Anunnaki to develop themselves as they awaited the return of the Twelfth Planet. These belief systems also include the idea that humans can advance only with the help of higher beings.

Complementing this seeking of higher realms is an almost universal sense among cultures that the very nature of incarnation reflects a "fall" from a more elevated state. Theologians have posited interpretations of a change in relationship to "God." Philosophers[12] have tried to explain such myths in metaphysical terms. One Earth-bound theory is much more straightforward: At some point there was a precipitous rupture in contact between Earthlings and more advanced colonizers. Sitchin, for example, describes the "fall" as the Anunnaki's expulsion of humans from their private domain (the Garden of Eden) into the wilds of nature.

From the perspective of this chapter, the local mind's sense of a "fall" may derive from its memory of the process of incarnation. That process involves consciousness giving up degrees of its freedom as it becomes enveloped in the less malleable *matenergy*. Personal development in the incarnated being may be energized by an impulse to regain those lost degrees of freedom, using all senses to realize the maximum possible cosmic awareness. If this is the case, then personal decisions and social choices that inhibit the individual being's reestablishment of its wholeness are violations of human purpose.

Altered States

Although universal consciousness is unidimensional in one sense, its nature permits multiple states of local awareness. Labeled in various ways by different people, the states are called *waking*, *altered*, and *sleeping* by Rosemary McMullen (writer, artist, and philosopher). She relates the three to creativity in the "objective systems" of language, art, and formal images. These states may fluctuate with internal cycles, as a result of chemical intervention and deliberate self-manipulation, or as a function of the manner in which the senses are employed. Regardless of the labels used—alert, dreaming, meditative, drugged, inspired, dazed, sleeping, focused, etc.—each one represents a position on the consciousness spectrum. The labels chosen for that spectrum depend on what a society considers useful. For example, Sanskrit has literally hundreds of words to describe different states of consciousness. One way of expressing the range of that gradient in terms familiar to Americans follows.*

States of Human Consciousness
• Sleep (preconscious)
• Drugged (semiconscious)
• Dream (subconscious)
• Awake (conscious)
• High (superconscious)
• Hallucination (unstable consciousness)
• Meditation (supraconsciousness)
• Channel (cosmic consciousness)

A brief look at these states[13] illustrates the variations possible in the range of consciousness common to all humans. One of the most obvious

* I believe the Principle of Correspondence predicts a comparable range of states in other cosmic beings. If such is the case, they are likely to experience various distortions in perceptions as humans do.

differentiations in the continuum involves the two poles of sleepfulness and wakefulness. As revealed earlier, science has not yet figured out the purpose of sleep and how it is precipitated. All human beings need some of it, with the length of time seemingly dependent on idiosyncratic variables. There are a few well-documented cases of people who sleep very little. Some scientists argue that a chemical shift causes the change in a conscious state, but it is also possible that the chemical change is only *reflective* of the shifting field of consciousness required for physical survival, i.e., the physical body serves the requirements of consciousness. For example, in many documented NDEs a person's indestructible consciousness is the most logical explanation as to why a clinically dead body returns to life.[14]

Ordinary human experience would indicate a being cannot be in both realms (asleep and awake) at the same time, unless an overall equilibrium can be achieved. An example of such a balance of the two states is the case of a meditating monk in the Himalayas who has reportedly not fallen into ordinary sleep for more than 70 years. If all awareness is solely devoted to keeping the five physical senses alert, a being is deprived of a full connection to the general field of consciousness. When such access is not possible through altered states, the only alternative is through sleep. The more we inhibit access to universal consciousness while awake, the more we may require sleep for the cellular level of recuperation that is necessary for balance among the three realms.

An artificially induced state very close to sleep is one of drugged semi-consciousness, when chemical substances neutralize all the sensory input channels and their synthesizing nervous system. (Richard Alpert/Ram Das reportedly gave up using LSD when he found that a meditating monk had clearer access to the *noumena* than his drug-induced states produced.) Several substances can accomplish this effect, which too ranges on a gradient from light to heavy. Its lighter phases also permit a merger with the dreaming state.

Dreams, flitting across periods of sleeping and waking, demonstrate the seamless nature of reality. They combine the data we access through the subtle and physical senses, making it possible to see, talk, touch, and also taste and smell in our dreams.

Dream use of the subtle senses makes it possible to receive information about distant events—problems experienced by a loved one, auspicious or catastrophic occurrences, and even unrelated incidents. Some persons experience precognitive dreams (of events to come) that could not possibly be stored in the brain's neurons. Any plausible theory of dreaming must account for all these characteristics of dreams. Inexplicably, in the light of human experience, current scientific hypotheses still postulate that the hodgepodge of images and feelings in dreams come from random electric misfires generated in the brain stem.

The awake state is assumed to be so well-known that it hardly gets discussion. Yet it, as all other states, is not uniform; there are shades of awakeness. Dreams called "daydreams" find their way into it as do reveries and other forms of musing. The highest level of consciousness appears to establish and maintain the necessary level of wakefulness, permitting deviations when its overall needs are met. Local consciousness is sometimes affected by external stimuli like loud noises or other focusing action, but the maintenance of the desired state requires deliberate focus, even if the intent is to remain unfocused.

A well-known example of a spontaneous shift from one waking dimension to another was experienced by Nikola Tesla in 1882. One evening while walking with a companion in Prague, Tesla received an instantaneous vision of an unknown machine—an alternating-current motor. After a brief trance-like pause, he reported that a vision had come to him in such detail that he could later, with his eidetic memory, draw a complete blueprint of a workable invention.

In the high state, in which the senses (physical and subtle) are expanded, one can play the range of all the lower states. The use of LSD or other hallucinogenic drugs can precipitate either a high (superconscious) or hallucinatory (unstable consciousness) state depending on the amount used and the general condition of the recipient.

The individual can control the input from general consciousness by choosing a topic or an area of focus. Through association, other topics with similar energy profiles then come into the person's field of awareness. In this way, related knowledge is brought to bear on the topic regardless of the state in which it was originally registered. This multilevel reception makes data from the dream and other states available to the waking state, and vice versa. (This overlap explains why the body's reactions, e.g., sexual or fearful, are the same whether the stimuli come in dream or awake states.) Some people learn to benefit simultaneously from the day dream and night dream states: with conscious access to the more subtle planes they provide themselves with a powerful resource for enhanced perspective and knowledge. They use this expanded information to increase understanding of self and others, and to perceive insights from the ageless *noumena*.

Within David Bohm's framework, where the manifest (phenomenal) and potential (noumenal) realms are the "explicate" and "implicate" orders, higher states of consciousness permit more creative play within the implicate order. Ranging at this level helps to identify and manifest more nuanced connections with other beings and to choose courses of action with a broader grasp of what is cosmically feasible. Developing the ability to conduct a lucid dream, for example, enables one to bring the realm of cosmic consciousness into ordinary awareness. Through the use of lucid dreams and other high volitional states (stimulated through meditation, ritual, or biofeedback), one gains access to the inner realm without the use of drugs.[15] Such access requires only the alignment of our conscious frequencies with the desired ideational field, which can be done by self-direction, as in remote viewing.

Some altered states are not always considered creative or benign. The label "schizophrenic" is used for some people who report perceiving a reality radically different from that of their associates, i.e., through "hallucinating." But others who have similar experiences and are not behaviorally impaired by them—shamans, religious leaders, and artists—are considered to be spiritual, prophetic, or visionary.

Perhaps the degree of social alienation in each instance is a function of how the behavior is initially defined by those making the judgment. Teachers and counselors may label one person's behavior as hallucinatory while initiates may label similar behavior in their spiritual guide as inspirational. Instead of encouraging shifts from one spectrum of consciousness to another (as group chants and musical patterns in many Hindu temples and Muslim mosques in India do), Western health professionals use a pharmacological fix for periods of unstable consciousness.

When schizophrenics are overwhelmed by the deluge of uncontrolled stimuli from the *energeia* and *noumena*, antipsychotic drugs (such as Clozapine, Haldol, or Thorazine) reduce the "aberrational" behavior through sedation (reducing the effectiveness of vision, memory, and movement). The patient is mentally dulled and has diminished physical faculties for expressions of any kind.[16] The effect is that one's state of consciousness is shifted from the higher level of hallucination to the lower, drugged one. The opposite approach would be more appropriate, developing thc human capacity to benefit from the full range of states, even when they are painful or frightening. Training in managing consciousness and support for its integration into education practices are needed for social progress.

Consciousness/Intelligence

In addition to the various states of local consciousness or mind, one can talk about its different levels of development or power. As more evidence of other types of cosmic beings is gathered, we project our own taxonomies onto the data. When faced with unknown beings, some people see angels, while others see devils.[17] For those focused on technology, other species of beings with advanced technology are perceived to be of superior intelligence, because intelligence is one indicator humans use to compare themselves to others. Are the nonhumans with whom we come into contact more intelligent, or have they just had more time to develop? Before leaping to conclusions, we should clarify a misconception about the nature of intelligence. General intelligence is not the same thing as IQ, or Intelligence Quotient. IQ is a measure of a person's performance on an artificially designed instrument that reflects how well certain skills have been developed.

General or cosmic intelligence, a function of the scope of one's awareness of internal and external circumstances, includes the emotional ability to clear and maintain the functioning of all senses. Human experience suggests that varying levels of such intelligence do exist, but the respective influences of inherent characteristics and external social conditions are not known. This "cosmic awareness quotient" appears to be the developmental result of progressive interaction and testing of *Selfhood* against *Otherness*, concepts that are more fully developed in Chapter 8.

While theorizing about extraplanetary levels of intelligence is still in the phase of interesting speculation, the issue of what causes different levels of intelligence among beings of the same species and between local species also remains speculative. If Cabrera has interpreted the glyptoliths on the Ica Stones correctly, different mental capacities existed among prehistoric inhabitants of Earth. According to Cabrera's deciphering, space-based beings manipulated the brain capacity of humanoids to create five or six levels of specialized beings. At the top of that hierarchy was "reflective scientific man." Was such a hierarchy established by genetic manipulation or social intimidation? Some evidence exists that, in the human case, the level of "cosmic awareness" is largely self-determined by the degree of subtle sense use one permits oneself.

Apparently all human babies have the capability to recognize the differences among all the sounds of any language. In other words, they are born fully conscious in this aspect of intelligence. By the age of seven months they begin to lose that capability for languages not used by others around them. So begins a socialization process that eliminates possibilities for experiencing the whole of consciousness through one important tool, language skills. Learning other languages later on, to some degree, reverses the process. The same narrowing and concentrating begins to take place in other components of general intelligence as cultural patterns of emotional expression, thought, and action are adopted and inculcated.

Just as socialization, a necessary function of incarnation, causes the loss of ability to hear certain sounds, it causes the loss of ability to perceive much of the *noumena* and the *energeia*. While intrinsic capacities to feel and think begin as universal endowments, expressions of them are slowly shaped, limiting subsequent understanding. From early childhood experiences often comes a "self-limiting prophecy," and incarnated beings become less intelligent or cosmically aware than they are capable of being.

Can one minimize this self-limiting process? What is the effect of developing and fine-tuning the subtle senses? Is general human intelligence just a scaled-down model (as "in the image") of ultimate cosmic consciousness? The experience of some would indicate that all can enhance and maintain the natural birthright of consciousness. For example, using clairvoyance, Edgar Cayce could diagnose a patient with only a name and address. An individual can deliberately access a potential scene from the future using only an hour and a date as reference points. Everyone has and can enhance these so-called paranormal capacities.

Continuum of Life

In the Chain of Being, humans have traditionally placed themselves in a "special" niche somewhere above the animals, but below the angels. How this actually came to pass remains a great puzzle, as we saw in Chapter 3. Now such assertions are becoming increasingly untenable. Able to extensively manipulate their physical environment, human beings undoubtedly have the most highly developed symbol and communication skills among current Earth animals.

If the ability to produce similar offspring implies a quantum level of self-ordering intelligence, many different levels of *beingness*, including all the plant and animal kingdoms, are capable of meeting this criterion. Applying information from earlier chapters, solar systems and larger elements of the universe may also have this reproductive potential. The criterion may also be met by microscopic particles that have been considered inert. Gaston Nassen, with a new type of microscope, observed heretofore unseen somatids[18] reproducing themselves. This and the evidence on biocommunication discussed earlier now leave us with no nondisputable dead/live or unconscious/conscious dividing lines.

Religion and science have historically refused to ascribe consciousness to the plant kingdom. The conventional wisdom was that plants could not move, feel, or see. But Cleve Backster and others have now disproved that hypothesis with research showing there is biocommunication within and between species.[19] Conversely, for several thousand years it was thought animals could not regenerate their parts as plants do. But in the eighteenth century Abraham Trembley started a line of scientific inquiry which proved they could. Now Robert Becker believes it is an inherent function in humans that may be expanded.[20]

Most people, except the biological determinists, still believe the fundamental distinction between the human species and the animal kingdom is the human possession of a soul or unique form of consciousness. Is this too only anthropomorphic arrogance? The evidence is mounting that we have oversold ourselves on this final indicator of our special status. Recent news articles reported on several conventional research projects that demonstrate other species not only enjoy conscious awareness, but can understand human efforts to communicate with them.

According to the various studies, dolphins receive instructions, coordinate their own behavior, and give a response showing they understand the human request. Upon seeing a hand signal, for example, they swim together, and then burst to the surface in concert with *behavior that relates to the signal*. A parrot identifies colors, counts, and expresses feeling with words. Chimpanzees have mastered a limited human-created symbolic language. (They do not have vocal cords to form words.) With these symbols they express preferences and indicate grammatical comprehension equal to a 30-month-old child. What pet owners have known experientially is now being confirmed in the scientific research laboratory: understanding between species is clearly a two-way street. Even insects are known to respond quickly and obviously to a thought request.

Some of the studies have revealed even deeper levels of communication among animals. Certain animals give others, including humans, warning of impending danger. Different species pass the emotions of depression and exhilaration among themselves. According to the experience of Native Americans, wild animals offer themselves as food to people dependent on them. These examples demonstrate that varying degrees of consciousness among species is another one of the gradients that characterize every aspect of the universe, and we are no longer sure where human beings fall on that gradient. Does interspecies communication refute the idea that a separate "soul" is unique to humans?

In addition to the falling of interspecies distinctions, all the old views of human limitations are rapidly fading. Workshops teach how to have out-of-body experiences. Untold numbers tell of near-death experiences. Thousands have reported encounters with humanoid-like, ephemeral beings. In this context, it is reasonable to conclude that humans are *no more divine* and *no less mortal* than any other species. We are simply and gloriously part of a cosmic epic of conscious life evolving among the stars, called to reach toward the heavens while grounded in this Earth. A Pueblo creation myth places us in this cosmic web.

In the beginning was Spider or Thought Mother who created the phenomenal world by spinning threads that became the universe. Then, from her own spider being, she spun threads to form the beings of people. For each being she spun a delicate thread that connected it to her own web. So, as the Spider Woman spins a web of destiny, each being is somehow woven into the pattern.

In such a cosmos, to know something of an individual is to know something of the Thought Mother or Logos from which we all come. Since consciousness has no finite boundaries that we can discern, logic would indicate that it flows in all directions from the powerful center—the womb from which we were all extruded. Thus, limited ideas about our nature limit every aspect of our emotional and physical reality. By ignoring our broader selves and our connections with all life, we consign ourselves to become the limited beings we think we are.

Dualistic thinking is part of the problem: the idea that five senses are paramount separates us from the power in the nonmaterial realm. We can bring the power of our subtle selves into play and change our physical selves. Using our subtle senses we can communicate more effectively with all beings. We can understand the origins of our emotional reactions and use them constructively rather than allowing ourselves to be abused by them. We can give and receive healing: more awaits the being who becomes fully aware of how his or her consciousness has access to the power of the undivided whole.

But like the misleading chimera of dualism, "free will" is another artificial construction of the rationalist, who attempts to identify something that is totally and willfully free—independent of desire, need, impulse or habit. Just as our senses (feminine receptivity) exist on a spectrum, so does our will (masculine expressiveness). No aspect of life is completely independent; conversely, volition is never completely blocked. Conscious beings must learn the inherent constraints on their freedom in order to be fully creative.

With a desire to assume our rightful place among the stars, we must master the long-term skills for a self-sustaining habitat on our home planet before we will be invited into the galactic neighborhood. No intelligent species would welcome or provide assistance to beings who foul their own nest and export destructive attitudes, behaviors, and technologies. If they understand the subtle energies, they will not wish to open themselves to unwarranted invasions by neophytes who have not learned to harness their thoughts and emotions.

Humans, through development of their meditative capacities and their material sciences, are now capable of traversing the quantum gap between mind and matter. While avoiding for the moment the abyss of nuclear self-destruction, they have peered through the lens of subatomic matter and perceived their own consciousness at play. Although still in a state of relative naiveté, they are equipped to probe for knowledge in all realms. Thus humans have passed the initiation rites into cosmic young adulthood, and can now engage with other conscious beings from common ground. They can aspire to sit in the halls of the galactic elders without fear of censure. The eventual fate of the species depends on how humanity uses the opportunities for maturation that are before it.

NOTES

1. O'Leary, Brian. *Exploring Inner and Outer Space* (North Atlantic Books: Berkeley, California, 1989).
2. Murphy, Michael. *Future of the Body* (J.P. Tarcher: Los Angeles, 1992).
3. Riversong, Michael. *New Science News* Summer 1993.
4. De Cuevas, John. "Mind, Brain and Behavior." *Harvard Magazine* Nov. 1994.
5. Rubik, Beverly. *Frontier Perspectives.* (The Center for Frontier Sciences, Temple University: Philadelphia, Pennsylvania, Fall/Winter 1991).
6. Zohar, Danah. *Quantum Self* (Quill/William Morrow: New York, 1990).
7. Jung, Carl. *Man and His Symbols* (Dell Publishing: New York, 1964).
8. Ornstein, Robert, and Richard F. Thompson. *The Amazing Brain* (Houghton Mifflin: Boston, 1991).
9. Dacher, Elliott S. *Intentional Healing: The Step-By-Step Guide to Mind/Body Healing* (Marlowe & Co: New York, 1996) and *Whole Healing:A Step-by-Step*

Guide to Reclaim Your Power to Heal (NAL-Dutton: New York, 1996).

10. Laberge, Stephen. *Lucid Dreaming* (Ballantine: New York, 1986).
11. Wills, Christopher. *The Runaway Brain* (Basic Books: New York, 1993).
12. Thompson, William I. *The Time Falling Bodies Take To Light* (St. Martin's Press: New York, 1981).
13. Electrical waves in the brain are associated with some of the different states of consciousness as follows: (1) wide awake, Beta waves at 13+ hertz (waves per second); (2) relaxed, Alpha at 8-12 hertz: (3) dreamlike, Theta at 4-8 hertz; and (4) sleep, Delta at less than 4 hertz. Many machines are sold to assist people in moving up and down these frequencies, although the Yogi or Sufi does so without technology.
14. Brinkley, Dannion. *Saved by the Light* (Harper Collins: New York, 1995).
15. Lapkoff, Cathleen J. "Adventures in Consciousness." *Fate Magazine* Jan. 1994.
16. Boodman, Sandra G. "New Hope for Schizophrenia." *Washington Post* 16 Feb. 1993.
17. Pacheco, Nelson, and Tommy Blann. *Unmasking the Enemy* (Bendan Press: Arlington, Virginia, 1994).
18. Bird, Christopher. *Persecution and Trial of Gaston Nassens* (H.J. Kramer, Inc.: Tiburon, California, 1991).
19. Tompkins, Peter, and Christopher Bird. *The Secret Life of Plants.*
20. Becker, Robert O., and Gary Selden. *Body Electric* (William Morrow: New York, 1985).

Chapter 8

Growth: Balancing Act

Individuals and their behavior are seen by Western science as products of physical forces that control our bodies and the universe. Western religions give no more leeway: we are seen as products of a distant god, subject to divine machinations. In either view there is relatively little scope for personal power and responsibility. Frontier science and traditional wisdom see human reality very differently. Human beings are nourished and constrained by that which is external, but inwardly they guide their own development and help shape the cosmos, thus, manifesting both aspects of the cause/effect polarity. This balancing act requires discovering one's talents and making the most of their power.

For the newly born infant there is little sense of difference between self and other, with distinctions learned only as others fail to respond to its desires. As it experiences some delay in need satisfaction—perhaps the mother's breast is not available when hunger arises—the baby begins to comprehend the distinction between its being and that of others. Experiences of early childhood establish the patterns of separation. The profile varies from culture to culture, but by early adolescence the boundaries are quite clear in all societies; in many, explicit rites of passage take place to mark the transitions to greater autonomy.

One's life is then spent managing the boundaries between self and other. Extremes of isolation and co-dependence are usually meant to be momentary: sometimes we need to pull our "shell" about us, and at other times we reach out like the multiarmed octopus. The process is a continuous balancing act. Americans are lulled into a false sense of autonomy by the myth of individualism and generally fail to appreciate the benefits and limitations of human interdependence. Paradoxically, individualism defines itself in relation to the others it shuns.

Chapter 7 dealt with *beingness* as an incarnation of consciousness within local space-time, from the perspective of how the individual derives from the multidimensional cosmos. This chapter looks at the three-level process of human socialization, the individual's developmental interaction with other multilevel beings who also have their own unique destinies. Danah Zohar, in *Quantum Self*, implicitly observing the Principle of Correspondence, compared the "thingness" of the particle to the self (personality) and its "wave nature" to the person's relationship to others. Using this quantum analogy, she notes that humans are therefore "waves" or "particles," depending on who is doing the observing. Employing the term *Selfhood* combines these two aspects, and indicates self is more than a static thing, since it involves the dynamic of relating to *Otherness*. (*Otherness,* as used here, includes other beings and the habitat.)

Continuing with Zohar's analogy, others are also *Selfhoods*, from their own perspectives. As in the subatomic microcosm, from one observation point the being is a *Selfhood* (particle) and from another it is *Otherness* (wave). Consequently, social reality is always a function of "relatingness," i.e., the outcome of reciprocal definitions along the *Selfhood-Otherness* continuum by equally powerful beings. We are beginning to recognize that each being has at least three sources of power (*phenomena, energeia, and noumena*) for this process of co-creation. Seeing the individual being in the integral, three-faceted universe recalls a lost sense of wholeness,[1] providing profound insights into human psychology that are missed in current theory.

Selfhood

To arrive at these insights, one starts with the three-faceted model where *beingness* comes from dynamic interplay between ideas and material forms, bonded by the cohesive forces of the *energeia*. The ability of what we call "living" forms to take in more ideas, energy, and matter, and transform them for their own use, makes self-development possible within certain limits. Categories of *beingness* below the threshold of organic life are also engaged in a constant interchange of matter and energy, but they appear to lack sufficient consciousness to make the finite quantum choices that are intrinsic to self-sustaining life, or *Selfhood*.

The evidence points to the possibility that at some level of group consciousness, humans literally hold their body cells together with a vision of who they are. On a larger scale, it is conceivable that conscious beings collectively maintain the integrity of the material universe through group mind. Although the power of creation is subject to certain constraints at the individual level, a single person can largely determine how well his or her body will perform as an athlete or dancer, or whether it will be ill or well. And the evidence is mounting that we also have a significant impact on other organic entities and our material surroundings. (Many anecdotes describe the ways computers, automobiles, and machines respond to changes in the user's moods.)

The ability of a being to self-monitor and self-maintain does not arise spontaneously. It derives from the moment of conception that joins together the three realms through the inherent principles* that support life in the universe. Not yet understanding the primal origin of this initiating consciousness, we can only accept and marvel at ourselves as the offspring of the Grand Couple. One great unknown is how many species, if any, have the power to take a primal idea and focus subtle energy enough to synthesize new life forms, without depending on biogenesis (the development of living organisms from other living organisms) or genetic manipulation or mutation. Is it only the Grand Couple?

---

* Though these principles are almost totally beyond an individual's influence, there is growing speculation that conscious beings acting in concert may have the power to influence the direction and degree of the universe's most constant constraints, such as the speed of light and gravity. [2]

According to Raimon Panikkar, "The ancient Greeks ... had already defined life (zoe) as *chronos tou einai,* the time of being. The very temporality of the universe manifests that it is alive; it has youth, maturity, old age, infirmities and even death. Zoe is set against thanatos, death.... Time is the very flow of being itself ... the peculiar way in which each thing lasts."[3]

For humans, *beingness* is consciousness in the here and now, in local space-time. Yet *beingness* precludes neither the simultaneous existence of different beings in other dimensions nor our access to them. In fact, the interconnectedness that is experienced through the subtle senses, OBE/NDE excursions, and other interdimensional communications reveals that local *beingness* cannot be severed from its internal, infinite source and context. In some manner, the cosmic umbilical cord is apparently never cut, even for a seemingly independent human incarnation for an Earthly lifetime.

Using the concept of Paul Tillich, there is a preexisting *ground of being* from which our *beingness* derives, but time is a part of this existence, a fourth dimension of the environment/circumstance that is a part of Self. While temporal life and *beingness* are conterminous, the human self is more than time and circumstance, and appears to have some degree of immortality.[4]

More and more evidence indicates that at birth humans bring pre-existing "knowingness" into this *beingness*. Recent research has shown that newborns less than an hour old can recognize a human face. Within 12 hours of birth they can distinguish their mother's voice. A father holds his daughter within seconds of her birth, establishing deep, mutual eye contact. But that which the father might label a moment of bonding, may in fact be a moment of mutual re-cognition. However, as the new being begins to focus more on the phenomenal realm, certain pathways to knowing appear to become constricted.

Noam Chomsky, noted MIT linguist, believes we are biologically prewired to learn language. As noted in Chapter 7, humans are born with the ability to distinguish among all the speech sounds in all languages, even artificial ones, but within a year the recognition skill narrows to those of the native language. Infants can even appreciate the emotional implication of words in a language they have never heard before. While most of the content of early learning appears to come from the environment, the process of comprehension and consolidation is innate. As infants, humans even appear to intuitively grasp physics, the difference between solid and holographic objects. Consciousness incarnate apparently learns about the externals of its incarnation with skills brought with it.

Such early demonstrations of knowledge come from a holistic, multi-sense understanding of this world that is a function of a being's total, eminent nature. Much of this multilevel knowledge becomes suppressed by the focus of consciousness required for life in the phenomenal world and by self-limiting cultural patterns. This brief survey of what we seem to understand leaves us with questions that require study by frontier researchers: What do newborns know? How is taking on human form self-limiting, and must it always carry the price of diminishing awareness? How can we maintain and enhance this broader access?

Many children report memories of specific earlier lifetimes, while others frequently have a sense of having already experienced a place or social situation. Much of the so-called fantasy life of children is surprisingly like material from other lives. However, society discourages such interpretations, and growing individuals soon forget their memories. Enough research has been done to convince many that the Self does bring earlier experiences to bear on its interpretation of and reactions to this life. It is a hypothesis worth wide-scale testing. If all of us in this incarnation could consciously open a basket of previous history, it would add rich ingredients to the picnic that occurs with our fellow travelers in this space-time.

Otherness

Why is *Otherness* so crucial to *Selfhood*? To a large extent, intraspecies relationships determine how we evolve in a lifetime. The universe has infinite species of beings and many forms of *beingness*, but all of them, like humans, live in communities of similar beings who mutually shape each other. To live more consciously and take fuller advantage of their capacities requires that humans better understand how they are differentiated from each other and, at the same time, how they are the same.

The individual's DNA "fingerprint" has only one chance in one trillion of having five matching sites on the estimated 100,000 genes in the 24 human chromosomes, yet every material particle in one's body has recently been in someone else or in the common resource base. At the same time, nothing we think or feel is totally hidden from others. Until humans recognize and feel these connections, we will not succeed in reducing alienation and its accompanying violence at all levels of society. Terrorism, ethnic violence, domestic abuse, and generalized anger and hostility are the results of failure to appreciate how all beings are part of each other. Crucially, these implications of "relatingness" do not apply solely to the physical realm: each should be considered in the context of the three-faceted model encompassed in the graphic of *Otherness* presented below.

<table>
<tr><th colspan="3">OTHERNESS</th></tr>
<tr><td rowspan="3">Subtle Energies
&
Emotions</td><td>Other Beings</td><td rowspan="3">Collective Consciousness
&
Data</td></tr>
<tr><td>Plants/Animals</td></tr>
<tr><td>Matenergy</td></tr>
</table>

One can choose to set the inner limits of *Selfhood*, but one cannot impose its surface limits on someone else. Therefore, the outer boundaries of *Selfhood* have to be negotiated with *Otherness*. The following graphic, used in psychology texts to demonstrate how one's mindset affects what one sees, also helps illustrate this negotiated process of self-definition. The *Selfhood-Otherness* boundary, while setting us apart from each other, also binds us together. This boundary is the space between the profile we have drawn for our self and the configuration drawn by others around us.

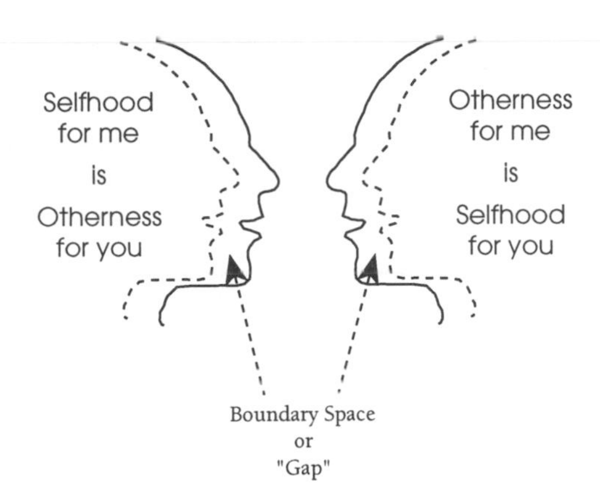

If one wants to change the inner boundary line, he or she can unilaterally pull in toward the core (dotted line), but if one wants to change the joint boundary *both* have to agree. One reason we have so many explicit roles (housewife, lawyer, teacher, parent, etc.) is that their conventions simplify the process of joint role definition, removing a large degree of ambiguity and negotiation. That the most fundamental core of a personality is jointly defined is illustrated by the fact that an artist's enduring creativity, by definition, must satisfy some continuing need or desire in others. Consequently, although human development involves some autodidactism, *Selfhood* is essentially a shared experience. To progress consciously is to grow with others, with each conscious being deciding where she or he fits on the spectrum of self-definition/other-directedness.

At one pole of that spectrum is the person who defines self almost exclusively in opposition to the behavior and boundaries of others. At the opposite pole is the person who becomes essentially what others expect. Carl Jung labeled this polarity introversion-extroversion: all life is an act of balancing between these two poles. Persons tending to counter-formation, in contrast to conformation, are considered to have "hard edges," a term that is descriptive of boundaries that counter the expectations of one's neighbors. Given the variety of human personalities, flexibility in taking different positions on that spectrum is necessary if one is to be effective in intimate relationships or successful in social organizations.

Otherness can be seen in terms of concentric circles, with the distance from self an indication of the person's or situation's relevance and importance.

Superimposed over such rings can be various patterns of relationship that define similarities and differences. Terms like "family," "kinship," "community," and "neighbors" usually fall in the former, while attributes like sex, race, and social classes may fall into the latter. (See the following section on differences.)

The importance of healthy joint boundaries and close circles becomes clear when a group's internal bonds of consciousness are externally challenged. Similarly, when an individual's relations with others are under stress, and that person has insufficient inner focus to maintain equilibrium, destructive behavior results. The psychic (energeial) and attitudinal (noumenal) forces released by covertly violent individuals or groups are as damaging to the society's field of consciousness as a bomb explosion is to the physical (phenomenal) world. Positive reciprocity must be maintained in all three domains. To understand the current wave of physical violence ripping apart community and nation, we can look to burn holes in the conscious connections among the members of these units. Large-scale engagement in emotional and ideational activities that reinforce the inner connections of cosmic beings could avoid such overt violence. Simply put, if we heal our inner life, the public one will change.

Peter Breggin, a psychiatrist who has studied human emotional problems by looking beyond the blinders of his profession, believes the psychology of an individual is a function of the psychology of the whole community.[5] Clearly grasping the reciprocating dynamic between *Selfhood* and *Otherness*, Breggin concludes that traditional psychiatry too frequently perpetuates a severance of the bonds of effective group consciousness. The use of drugs, electroshock, and surgery, in his view, damages the connecting web of consciousness rather than heals it. He is correct: the whole social organism contributes to the creation of individual problems (this is nowhere more evident than in the current internecine warfare among youth gangs), and only the whole can heal them. Healing of communities starts with a conscious choice to expand our own sense of relatedness. Hillary Rodham Clinton's book *It Takes a Village* is based on such an understanding.

Choice

Otherness begins at the point along the *Selfhood-Otherness* spectrum where an individual's direct control ends. In this regard, the human ability to maneuver in relation to others in the subtle realms is in many respects analogous to the capabilities of a cellular telephone: our movements are practically unimpeded; we send signals simultaneously with others; any one of us can contact any other on the planet if both parties desire; but, each has the choice to answer the call or not.

Communication employing the subtle senses works in the same way: all it takes is for two beings to open the channel. We can tune in on the broadcasts of all others, and receive the busy signal of those who are choosing not to be open. The scope of eventual communication depends on the intent and skill of the initiator (caller) and the willingness of other beings to engage (pick up the phone). The human's ability to exercise that choice provides the potential for access through the subtle senses to any part and, ultimately, to the whole universe, through the "cosmic internet."

The more expansive people want their field of knowledge to be, the more they must engage in mutually beneficial reciprocity with others. Conscious, positive understanding is infinite when we open ourselves in constructive engagement with other beings. Respect, like love, is infinite: the more one gives away, the more one gets. The boundaries between any being and *Otherness* are dynamic, continually changing as a function of the conscious will and the choices of the beings involved.

The key to happiness and growth is finding an appropriate balance between *Selfhood* and *Otherness*. Defining and realizing that balance is the ultimate aim of conscious beings. To be truly harmonious, one does not give up individuality, but enhances it with the help of others. Living according to one's core needs and allowing others to do the same recognizes that the individual's desires are only one factor in the algebra of cosmic consciousness: as any student knows, both sides of the equation have to balance for the result to be workable.

Supporting Others

Others are essential for developing *Selfhood*. The newborn is mentally awakened by the more conscious beings around it. They draw its attention with playfulness and soothe it with nurturing, helping the new being learn to balance between reception and expression, between venturing out and drawing into its own unique pattern. Wise elders offer both stimulation and repose, but allow the infant to learn how to select between them at will. The early experiences either reinforce innate tendencies to openness or negate them, shaping a being's life-long approach to the expansion of consciousness.

The insights derived from today's frontier research into consciousness have implications for our current theories about social behavior, particularly so-called antisocial behavior. Humans literally do not think or do anything alone: no thought or action occurs in isolation from other beings and our planetary habitat.[6] Consequently, our fate is to function like members of an improvizational jazz ensemble, playing the instruments we have in concert with other members, agreeing in advance to a set of chord changes, then making up parts in relation to each other in progression through the set. When the rhythms and harmonies "work," the outcome is music; when any element is off, the result is discordant sounds. Therefore, any accurate analysis of discordant behavior starts with the self, as one "musician" in a multifaceted relationship with the whole group.

Any interaction between beings involves energy. Even mental or verbal exchanges focus energy that in turn affects another's performance. If we unthinkingly project our shadows (the parts of ourselves that we ignore) on others, we project a burden onto their role playing. Public leaders are, to a large degree, either weighed down or uplifted by the energies focused upon them by others. We can assist in bringing about more positive behavior in leaders by projecting constructive energy fields into the public domain. The place to start is with our most intimate relations, whence it will ripple out in larger circles.

Individuals tend to perceive reality through their constructs of *Selfhood*, and then interpret *Otherness* through projecting that reality. If our sense of reality is to be more than a subjective attribution, there has to be confirmation from another being. Since the reality of *Otherness* is a collaborative projection of respective *Selfhoods*, so-called objective reality can only be a function of intersubjective agreement between *Selfhoods*. Recognizing this subjective and changing nature of external reality, useful science, like effective politics, involves continuous testing, modifications of assumptions, and repeated validations of what we think we know.

Form as Costume

All species of beings encompass variations in the three realms. States of *beingness* can include differing profiles of frequencies, amplitudes, and wavelengths on each spectrum. Such differing combinations could account for the existence of beings quite unlike us in outward appearances. For example, when people use the term "beings of light" in describing contacts in NDE or subtle sense perceptions, they are obviously referring to more amorphous forms of *beingness*. The range of feasible combinations may be almost infinite, including ghosts, aliens, and angels. Humanity should not be surprised at the varieties of form, ability and character exhibited by the cosmic siblings who will be encountered by coming generations. An analysis follows of the implications of some "real" differences and some "real" implications of differences that are only a matter of social convention.

Problems arise when humans alienate themselves from others through fixing categories of "outsiders" versus "us." These judgments may involve distinctions in race, language, religion, sex, politics, economic status, dialect, life style, eating habits, or any one of scores of other categories. Every one of these divisive characteristics has, at some time and place, led to hostile behavior against the group perceived as "outsiders." A recent poignant example is the brutality of the "ethnic cleansing" perpetrated by Serbs against Slavic Muslims and by Serbs and Croats against each other in the former Yugoslavia, where centuries of mutual hostilities-in-consciousness underlie the abuses of the 1990s. In that region, where there are few discernible physical and behavioral differences between neighbors, religion alone is used as a pretext to commit atrocities of pillage, rape, and murder. This behavior, and examples like it in all parts of the world, can be explained in terms of dynamics among conscious beings.

Violence against "outsiders" is the result of a two-step process. The first step is labeling of others as sufficiently different to cause mental alienation: This deliberate act of conscious classification results in the distortion of both the *matenergy* and subtle energy flows in the direction of the "outsiders." The recipient's experience is invasive subliminal turbulence. This two-way stress occurs on a subtle level, even if the physical manifestations are not obvious.

The second step is a decision to take action against the "outsider" in order to relieve the stress and return to equilibrium. Reaching this conclusion gives one the "permission" to rampage. Attacks, looting, rapes, murder and other forms of mayhem, whether verbal or economic, result in a temporary dissipation of the clogged energy flows. There is an orgiastic release at one or more levels, returning the violators temporarily to a sense of ease and harmony.

But since the fundamental problem—the conscious denial of interconnectedness—keeps the subtle energy channels dysfunctional, the pressure soon rebuilds and the same illusory remedy is tried again. This kind of "senseless" violence occurs only among conscious beings with minds capable of assigning labels or calling names. (The *Course in Miracles* says only a stranger can evoke fear; call him brother and he comes into your heart.)

This model of the origins of violent behavior does not support the thesis that such negative outbursts are caused by genetic and chemical imbalances, or environmental pressures and social disadvantage or misfortune, although these factors can influence the degree of dysfunction. Thus violence from within the homes of the "best in society" should not surprise us. Only by honoring the common bonds of consciousness, through an open acceptance of the reciprocal mental, energetic, and physical exchanges, can conscious beings avoid the explosion of internecine violence. Effective solutions may include conscious reopenings of the flows between beings by removal of the divisive label or involvement in games (some as simple as midnight basketball) and rituals that dislodge the thwarted energetic exchanges. Increasing external pressure or controls will only exaggerate the imbalance. So-called wars on behaviors that deviate from group norms have an opposite effect from that desired.

Overtly destructive behaviors differ from more subtle conflicts of ideas and emotions only in degree There is less societal concern about the milder end of the behavior-emotion-idea spectrum because of the lack of understanding of the indirect physical impact of conflicts in the subtle realm. Because the dynamics of the subtle soon become manifest in the physical, we need to be as concerned about how people relate on the inner realms as we are about the way we engage on a physical level. Given the isomorphic nature of the two, we can begin to cope with the overt problems of conflict by experimenting with how we relate to others on a verbal and feeling basis. The way to "civilize" undesirable behaviors is not to ban or suppress them, but to transmute the negative subtle energy involved into constructive forms. A simple exercise can prove the point.

The next time you encounter a minor frustration with anyone, try perceiving that person as a cosmic being like yourself. Recognize, through your inner eye, that she or he has the same essential attributes and powers you have in yourself. Think of their glowing *Selfhood*, shaped by consciousness and infused with the same cosmic energy coursing through your chakras. Sense the pulsations of that energy channeled by their unique patterns of incarnation, derived from the same reservoir of potentialities from which you selected your differing qualities, skills, and aspirations. Express the rising pulse of energy as a compliment or boost for some specific trait of the other. Then sense the shift that takes place in your internal energy balance and overt behaviors. Reflect on the reasons for the shift. (One can apply the same principle to diseased cells in the body.)

The way to heal the divisions caused by the labels listed earlier is to review each one and observe that the differences so immediately important to us are of little significance on the cosmic scale, which is our most important plane of existence. Below is an analysis of the phenomenon of race that reveals the lack of justification for its use in fomenting social divisiveness. Its use for psychic alienation is no more appropriate than gender, linguistic, religious, or any other such differences.

Currently we have not been able to explain satisfactorily why the human species has different races—another huge gap in scientific knowledge we choose to ignore. Theology, with its *ad hominem* assertions, attributes racial differences to an act of god. Evolutionists offer two equally unsatisfactory theories. One is that the three basic races evolved from different subspecies in separate locations. Such speculation is undermined by the DNA evidence that traces all humankind back to a common gene pool. The other evolutionist view is that mutations in the genes shaping skin color were a response to different climates. There are two problems with this hypothesis. First, no evidence exists of gene mutation due to such environmental factors. Second, racial differences also include varying facial features and body types. Even if we accept the argument that prolonged exposure to the sun results in a race of dark-skinned people, variations in light do not account for the other racial differences. This hypothesis does not hold up for another reason. There is no evidence to preclude the presence of light-skinned peoples in hot, humid or desert regions during the time scale presumed necessary for evolution.

Although no definitive evidence of the origins of human races has emerged, at least one explanation—no less documented than other theories—has the merits of being consistent with what is now known to be feasible. The views of Sitchin and Cabrera offer a possible explanation as to why all human beings share the same basic gene pool. Ancient gene-splicers may have performed operations on slightly different families of primates, resulting in the physical differences we now label as racial. Or if various individuals or groups of extraterrestrials performed such genetic experiments somewhat independently, they could have, by chance or design, altered a few patterns that resulted in racial differences. In either case, racial differences make no contribution to consciousness and behavior, and remain substantively irrelevant among cosmic beings. They are analogous to differences in the colors of automobiles—immaterial to performance and worth.

Plants and animals have significant differences in relation to humans and more advanced beings, but they too are part of the indivisible web of cosmic consciousness. The way we choose to relate to them, just as the way in which we relate to other advanced beings, joins the never-ending circle of cause and effect. Cruelty expressed toward any species by individuals or collectives will come back to haunt the whole and all of its constituents, while nurturance redounds to everyone's benefit.

Cosmic Family

Many living today have had dramatic experiences of contacts with nonhuman members of the cosmic family. Stories from most cultures indicate there have been similar contacts with beings from other realms throughout history. Examples of this aspect of *Otherness* are given in several sections of this book, but an illustrative review here reminds us of the *Otherness* from outside our own mundane realm that has historically shaped the lives of Earth beings.

Many accounts from the Near East of nonhuman beings with advanced knowledge have interesting parallels in Somerset, a county of England. The name itself may have been derived from Sumeria (land of summer and perpetual youth) around 3,000-4,000 B.C.E. The Druids knew of the extraterrestrial Annunaki zodiac and Sumerian astronomy. The name for King Arthur may have derived from Arcturus, the Great Bear of the Big Dipper and the brightest northern star. Uncannily similar to the twelve tribes of Israel and the twelve tribes of Delphi, Glastonbury was divided into twelve hides (1,440 acres each), which recall the twelve houses of the zodiac. Ancient Delphi and Glastonbury, sited on springs with magic waters, were considered communications links to more knowledgeable beings.[7]

In the original Celtic language, Iniswitrin, or Glastonbury, meant "the isle of crystal." It was known as the location of a crystal palace (off-planet vehicle) in a fairy fort. In fact, all the area was known as the land of the fairies (nonhuman beings). Glastonbury was reportedly ruled by Gwynn ap Nudd, a being who came from deep underground. As the story goes, St. Collen confronted Gwynn in the fairy fort and then later disappeared for a time in his palace cum space vehicle. In a parallel English legend, Guinevere, like Persephone of Greek legend, was abducted to another world. These stories are not unlike some told by modern-era abductees.

Whoever the ancient others were, humans, co-existing with them in many locales, saw these beings as natural members of their cosmic family. Popular movie classics like *Star Trek* and *ET*, television programs on angels, and modern fairy tales help keep alive these ideas of an expanded cosmic family, until we are able to be more officially public about its existence. In the meantime, we are left with isolated and disparate individual and small-group reports of communications with other members of our cosmic family. Without a generally accepted theory to explain these experiences, society is subjected to much misinformation and distortion of meaning in these interdimensional contacts.

Humans currently engaged in various forms of communication with beings from other realms should keep in mind that such beings are shaped by the intellectual and cultural histories of whatever galaxy or dimension in which they exist. Therefore, the reception of a message from one of those beings via channeling, direct contact, or other media does not mean the human recipient has been given a cosmic truth. Our human tendency has been to idolize the recipients of such communications and designate them a class of unquestioned arbiters of a divine wisdom. We have seen them as our prophets, our saviors, or our gurus.

Communications with other realms should be taken for what they are: expressions of individual beings or groups sharing their own experience with Earthlings. Reading, hearing, or being a channel for such communications sets no one apart: we are all channels, exchanging data continuously with beings and dimensions beyond ordinary reality. There is a cosmic ordinariness to all this (the multidimensionality of conscious *beingness*, the multiplicity of messages and channels, and the inescapable links of any being with all others) that should make easy our efforts to maintain balance and avoid overly dramatizing any particular message or interpretation. Maintaining a cosmically-grounded perspective is required.

A 1996 *Newsweek* poll revealed 48 percent of Americans believe UFOs are real, and 29 percent think official contact has been made with aliens. Although increasing numbers of humans are becoming directly aware of their membership in a cosmic family, for many that has been a threatening interbeing experience.

A 1991 Roper Organization survey, published as *Unusual Personal Experiences*,[8] came to the incredible conclusion that, based on its sampling techniques, perhaps two percent of the American population—more than five million people—have had experiences consistent with a UFO abduction history. (Note the term "abductees" excludes contactees—those who have had an encounter, but have not been taken aboard a craft for physical and mental examinations.) Roper used as survey criteria subjective impressions given in response to questions that had been administered earlier to 500 "abductees" under hypnosis. These responses were from case studies of two researchers: Budd Hopkins and David Jacobs, widely known counselors and authors of abduction-based books. Five types of their cases' unusual experiences were codified into yes/no questions. Using this basis, Roper asked a random sample of respondents if they had: (1) awakened, paralyzed with a sense of a presence in the room, (2) believed he or she could not account for an hour or more of lost time, (3) felt as if flying through the air, (4) seen unusual lights in a room, or (5) discovered scars on the body without a memory of how they got there. Hopkins and Jacobs concluded that persons who answered yes to at least four of these questions were probably abductees.

Extrapolating from the American sample's responses to the world at large would indicate more than 100 million living abductees. Numbers of this magnitude dwarf even the most speculative estimates of UFO activity. Could the polling instruments be faulty, or is the phenomenon of shared memories another explanation for such large numbers? Although UFOs have been reported for centuries and widely documented for the last 50 years, there were very few detailed abductee reports until best-selling books popularized the ET experience. The early stimuli were the Betty and Barney Hill 1961 case; 1973 national publicity of the Pascagoula, Mississippi, case; a 1975 film of the Hill case; and the 1981 book *Missing Time* by Budd Hopkins. Since then, the numbers of people claiming to have such experiences increases exponentially with each new sensational best seller.

Though the memories of many alleged abductees may be attributed to so much publicity, for the purpose of understanding how humans are embedded in a larger reality, evidence of shared memories is almost as important as contacts with other beings.

Other Influences

As recognition of a larger reality causes us to redefine the nature of ourselves in relation to all species, it also modifies the manner in which we relate to the dynamics and content of all the *noumena*. Beyond our ties with other beings, we are in constant communication with varying expressions of consciousness emanating from both organic and apparently inanimate sources, from something as complex as geometrical crop formations in the grain fields of England to flashes of intuition about Internal Revenue Service (IRS) regulations.* Both examples represent different ends of the spectrum, from possible bilateral communication between species to unilateral forays of a local mind into a specific database. Both appear to illustrate interactions of human consciousness with the external environment for our benefit. Simpler ones, like finding water or tapping into the arcana of tax regulations, provide help for a moment.

Some believe more complex interactions, e.g., human consciousness influencing designs in crops, may point toward salvation for our species and other life on the planet, somewhat analogous to the dreams Moses had that led to the survival of a nation. The capacities of individuals and groups to access universal information through the subtle senses bring to our attention messages that illumine dangers or expand our awareness, pointing the way to constructive change. Though in the case of crop circles, we are not sure an interactive process involving human consciousness is at work. Efforts to decipher what is actually happening may unlock unimagined human powers.

In August 1991, a crop formation resembling a human brain was found in a field of ripe grain near Froxfield, England.[9] Some speculate that this pattern may be a warning from our collective inner selves to reform our selfish interaction with *Otherness*. Along the same vein, the repetitive snail formations in the crop of 1992 are seen as warnings that we are reacting too slowly—a snail's pace is not quick enough. Perhaps at some level a critical mass of people sense the impact of our behavior on the material world and know corrective action is required. They may then manifest anxiety signals that alert collective thinking in the *noumena*, with the result being subtle energy focused in the phenomenal world as crop formations. Even where hoaxers are involved, they may be unconsciously channeling messages from the collective consciousness.

---

* While working on both this text and my tax return one day, I let my mind wander as I contemplated how to pay an amount I had not anticipated. Suddenly I felt the urge to check a particular section of the regulations. Sure enough, a deduction of which I had not been aware was available for the problem area.

A crop formation, appearing in a field near Cambridge, England, in August, 1991, clearly replicated the form of the Mandelbrot Set (shown below) generated in the 1960s by IBM's Benoit Mandelbrot. Through mathematics and computer graphics, Mandelbrot was able to depict the relationship of chaos (the massa confusa of the Hermetics) to form. The fall into matter (a ripened field of wheat) by a form (the Mandelbrot pattern) may demonstrate to us the potential that focused human consciousness has for creative interventions in the realm of *matenergy*. It may be a contemporary example of how the Biblical *logos* could have moved on the face of the deep and transformed chaos into order. Both the process (human concern possibly manifested through plants) and the content (the inherent order in apparent chaos) deserve our research.

As of this writing, crop circle researchers have not been able to develop foolproof criteria to distinguish an authentic crop formation from a hoaxed one. There is convincing evidence that many of the formations were the creative inventions of people wishing for various reasons to obfuscate the truth. But to ignore all of them because of some known hoaxes is to lose a precious opportunity for insight into the living cosmos. Many formations are associated with lights and shapes in the sky usually identified with advanced beings, but the UFOs may have been attracted to the phenomenon and not creators of it. Equally plausible is the hypothesis that the signals are human warnings to ourselves through this medium, with a channel of communication from the human brain to the plant cell, like that reported by Cleve Backster and others.

In the summer of 1992, in an attempt to test the hypothesis that we can interact with the crops through the medium of consciousness, I designed an experiment to determine if we could influence the formation of particular shapes in the grain fields of a selected area in England.[10] Around midnight, a group meditated at length on a secret common image (a bow tie) on top of Silbury Hill. Some days later, a formation—subsequently claimed by hoaxers—appeared in the field where we had focused our vision. A bow tie was included among the several designs composing the formation. Our experience was not unlike the sequence involving a letter Martyn Hughes sent to the *New Scientist* in August, 1991. In it Hughes asked, "How long before we see a Mandelbrot Set (in a crop formation)?" and a design like that shown above appeared a few days later.

A connection between human consciousness and the so-called crop circles is only illustrative of the many possible interactions between individuals or groups and the vibrant *Otherness* of our cosmic habitat. Industrial society has forgotten the tradition of such human interaction as manifested in Native American rain dances, spring fertility rites, harvest festivals, hunting prayers, etc. If the human mind can influence the pattern of falls by small steel balls in a cold mechanical cabinet,* what is the potential for a powerful exchange of influence between us and other living species? If we can engage particles in a dance into and out of form, why is it not possible to consciously obtain reciprocal responses from weather patterns and other aspects of our environment?

---

* Such are Robert Jahn and Brenda Dunne's findings in the Princeton Engineering Anomalies Research (PEAR) laboratory at Princeton University. [11]

Testing Theories

In the discussion of *Selfhood* in relation to *Otherness*, two hypotheses have been either implicit or explicit: (1) Since no single being has a corner on wisdom, anyone and everyone can have access to any and all knowledge. (2) One can validate the source and content of specific information received by another being through one's own personal experience. This means no one has to depend on an untested other as the arbiter or interpreter of esoteric knowledge. Any *Selfhood* can confirm the distinction between personal experience and shared memories. The following paragraphs illustrate ways to think about personal testing of these hypotheses.

With regard to the first hypothesis, the next big step for us as Solarians is to give up the myth that has kept us passive children—the belief that there is a separate inner being, or an external being, that is a magical source of control. Father Gods, Holy Spirits, and Atmans/Brahmans, or the Grand Couple for that matter, have not revealed themselves as being involved in human life in any confirmable way. In all the world's belief systems, attributions of such involvement are only untested assumptions. These have served humanity as crutches for evading responsibility for conscious living.

Mohammed, Jesus, Edgar Cayce, Sai Baba, Rudolf Steiner, Albert Einstein, and Nikola Tesla all benefited from communication with other realms, but so have millions of others. The messages or reputed revelations of such individuals are inherently no more sacred than many others. The value of each insight depends on the source (How wise was the nonhuman originator?) and its relevance to Earth reality. No longer can we evade responsibility for testing alleged universal principles by labeling them "commandments" attributed to a "god." Received "wisdom" cannot be considered definitive just because it appears to be from another realm. That a message is perceived to be from an "angel" does not relieve humans of the responsibility to validate it on this plane. Data from the *noumena* merits our consideration only because it may be based on more experience and perhaps has stood the test of time in other realms. If it has, then it can stand the test of our scrutiny and efforts at confirmation.

The earlier discussion of the Roper poll on alleged abductions by nonhuman beings in our cosmic family is also relevant to the second hypothesis mentioned above. Since the end of WWII, the numbers of reported cases of abductions have increased at an almost geometric rate. Why should this be the case when there has not been an equal rate of increase in the discovery of actual physical evidence of alien visits? Perhaps there is another explanation: many people may be sharing the memories of one individual's experience (cryptomnesia). (Keep in mind that shared memories are not the same as false memories.)

Many current researchers have engaged in a process of circular reasoning that cannot be considered external validation of such a mass increase in contacts. Without corroborating physical evidence, it is just as plausible to conclude people are reporting experiences that are not their own. Some discussion of that process is merited here because the phenomenon is central to fully understanding our Solarian Legacy.

There are three levels of abduction reports: (1) recall bolstered by clear physical evidence; (2) memory without loss of consciousness; and (3) lost memories recovered from the "unconscious." Abduction researchers have identified many cases that were self-contained, phenomenal experiences with a material record. They have also described the experience of persons who report conscious awareness of abduction, but exhibit no physical symptoms and have no other objective evidence. The problem arises when the researchers treat similar reports of unconscious memories elicited by hypnosis as confirmations of ordinary reality. The withholding of certain details from public reports for validation purposes does not guarantee the absence of contamination, since data is exchanged through the subtle senses. Therefore, the correlation of "memories" should not be considered a validation of actual experience.

In the self-contained research protocol (shown on the next page), there is no provision for external validation: as long as they are similar, memory reports are assumed to be based in actual events. Psychosomatic and psychological symptoms that could have third-party-memory explanations are therefore interpreted as confirmation of the hypothesized actual experience. The loop turns back on itself: repetition by more and more people of similar reports is taken as confirmation of actual events.

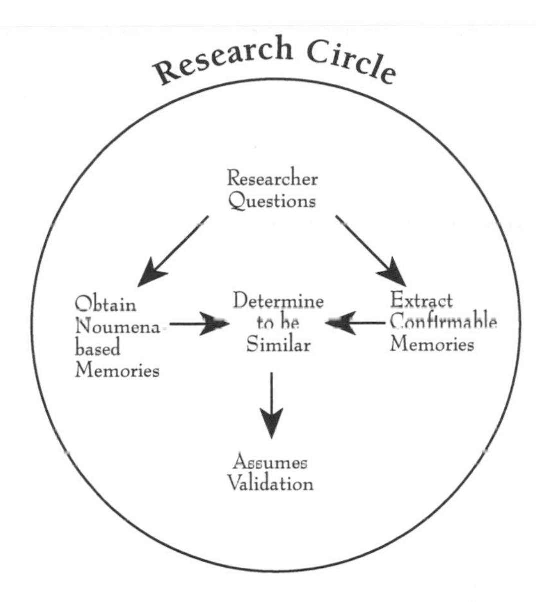

In order to be convincing, analysis of abduction reports must take into account the dynamics of memory and the existence of a common pool of ET encounter images, accessible to anyone who is attuned to the same vibration as the participants in the originating event. Individuals can tap into this common pool through dreams, fantasies, meditation, telepathic efforts, and hypnosis. With self-conscious checking they can discern the difference between what is genuinely of their own experience and what is not, but such filtering is not possible in a hypnotic state. Therefore, the researcher must build in external controls. If the researcher takes the material literally and tells the subject's waking mind that jointly they have accessed reality, a susceptible person will accept it as truth without subjecting it to thoughtful external validation.

The result of the above-described flawed process is that information the self obtains from the *noumena*—information attached to the ordinary experience of another—is labeled as one's own and acted on accordingly. The boundaries between *Selfhood* and *Otherness* have been reshaped by unwitting outsiders who have contributed to the delusions of personalized experience. This caution applies to any area of experience, not only to ET-related events. In the indivisible field of *Otherness* of which we are each a part, it is crucial to personal integrity to maintain an awareness of the reality of *one's own* local space-time experiences, even if it involves seeking the help of others to confirm perceptions.

Thoughtful Interdependence

Conscious living is based in a full awareness, not only of the invisible web that binds all beings to each other, but of the many ways in which it connects us. Every "event" between *Selfhood* and *Otherness* is no more than a definitional point on a continual flow simultaneously involving all three facets of all beings. The material (*phenomena*), emotional (*energeia*), and informational (*noumena*) bodies of all beings are engaged without ceasing; the various vibrations change rates according to their natural rhythms. Therefore, when we stop to examine or discuss a specific event, its isolation from the ongoing flow is artificial: its reality is embedded in a larger context that must be kept in mind.

For example, in the case of a "shared-memory abductee," where *another's reality* is made part of *one's own reality* by appropriating the other's memories, the borrower subconsciously incorporates (take into one's own corpus) not only the other's ideas, but also their emotions and physical reactions. Unless one exercises conscious choices and personal validation, a power inherent in all humans, responsibility is abdicated.

Therefore, the dance between *Selfhood* and *Otherness* is always the context for choice, the chance to be a leader or follower. One either primarily creates original experiences or shares in those created by others. Any moment of behavior is usually a combination of both, but here the focus is on the degree to which one is dominant over the other, to illustrate the nature of individual choice. Given that interaction is constant, the key to successful living is the balancing of these polarities with rhythmic, small pendulum-like swings without too much emphasis on one extreme or the other. The rhythmic dynamic is manifest in changing gender cycles of giving and taking, passing through a neutral phase in between, like the reversing of electrical charges or the phase changes of a magnet. In the Romantic's vocabulary, we are easily both the lover and the beloved, shifting positions at will. The polar impulses to merge and to separate have equal force, with the latter totally dominant at birth and the former at death. How we play the yin and yang aspects of many different roles in the changing rhythms of life will determine the nature of the being we become.

As we move beyond the biological concept of humans, with a discipline of psychology that has had to invent unconfirmable subcomponents of personalities to account for anomalous behaviors, we will approach psychological theory building and research from the perspective of at least three aspects in an indivisible field. The psychology of the future will start with the study of indivisible consciousness, then move on to local minds (incarnate beings) and how they are integrated into the whole. Practical psychology will focus on self-directed management of personal awareness and its interdependence with *Otherness*.

The actual nature of interdependence of any one facet of *beingness* with all *Otherness* is difficult to pin down. Interdependence is like the force of love between beings who are never completely without its presence, yet awareness of it rises when they are in physical proximity or exercising focus. Moments of awareness of the unbroken web of life-shaping connections may rise and fall in a serendipitous manner, but these two-way flows do not occur by chance. They are always operative and available for conscious use. With the Principle of Cause and Effect we can feed ideas into the *noumena* and observe their effects come back to us. Conversely, as we express the need for information or assistance, it comes to us. The surprise should be not that it comes, but that the source (*Otherness* over which we think we have no control) actually cooperates with us. As we "feed" others ideas and subtle energies, they in turn "feed" us. There is no better evidence of the human power of co-creation.

After I became interested in noting such "coincidences," *my ideas* on Third World economic development, still in draft for future columns I was writing for the *Washington International*, also surfaced in other organizations or publications with no direct access to my work. As I was following a path of psychic development at the time, I became impressed by my powers of remote influence. Then it dawned on me, it was a two-way street: these ideas were literally "in the air," "floated" by some unknown being. Given the correspondence between realms, we can find subtle energies or emotional fields in the air to serve us as easily as we can find ideas.

Some in the New Age movement tout the idea that mind, exercised through visualization and positive thinking, can unilaterally influence, cause, induce, or persuade matter to behave differently than it would if left to its own processes. However, this hypothesis is based, erroneously, on the implicit assumption that mind and matter are separate, requiring the former to communicate with, transmit to, or interact in some mechanical way with the latter. This concept is, nevertheless, an advance over the conventional scientific paradigm in that it presupposes the local mind as the more basic and the more powerful of the two. But in all other respects, it is a variation on Cartesian dualism. Experience teaches us that science *and* the New Age movement can benefit from recognizing the concept of interdependence, in which local mind does not exist without matter, or matter without mind.

The Hermetic Principles provide for no clear dichotomies, either at the microcosmic or social levels. As we have seen, the proof of such unity can be found in modern physics, where quantum theory suggests an undivided context surrounding any event or phenomenon. Even a scientific experiment cannot be divided into separate components; the experimenter as observer helps shape the outcomes at the same time she or he is constrained by its boundaries. One personality cannot exist independently of other persons. Our vocabulary makes it difficult to describe the process because we define as distinct such states as brutality, love, hunger, heat, solitude, maturity, etc., when they are really points on many continua that form facets on the same cosmic diamond. We need new metaphors for these interactive processes and states.

The image of the dowser is a useful metaphor for grasping how one can obtain and use noumenal information in the material realm. Dowsing generally implies the process of locating water or lost or other material objects through an inner sensing that is reflected in the movement of rods held in the dowser's hands. But dowsing can mean extending the field of a person's conscious awareness to access any data that already exist; the metascientific dowser opens herself to the multilayered vibrations from the desired object, person, event, or energy field. Dowsers of ideas can tune in on others' thoughts with their divining mental antenna. Reciprocity occurs when such dowsers—which we all are—have their own ideas picked up by others.

Another useful metaphor is the gardener who plants seeds, nurtures, and protects, yet depends on elements outside of self to determine if the right temperature, light, and moisture are provided for growth and maturation. Cosmic gardeners can take the initiative, respecting the relevant principles, but must be in congruence with the vast *Otherness* to ultimately enjoy the fruits of their labors.

To illustrate just how difficult it is to convey the essence of interdependence, even the Hermetic idea of cause and effect may seem to imply separateness and an independent flow of force. However, a cause and effect relationship in a totally unified field is different from one in a system of segmented parts. In an integral universe it is impossible to isolate a single, independent, causal relationship. While we do not have an understanding of how the interrelationships work, the subtle sense mechanisms earlier hypothesized merit further study. They are a starting point for understanding what is likely to be multiple flows of interactions at work.

At this point, we do not know through what media the subtle senses work: humans take advantage of them as people enjoy television, without knowledge of the electromagnetic spectrum. Humans can enhance their subtle abilities in the dark, so to speak, but great leaps are likely when we grasp the principles involved. Two basic assumptions for research are faster-than-light, but measurable links parallel to the electromagnetic spectrum (suggested in Chapter 4) and intangible connections involving no measurable force or interaction. John S. Bell is among theorists who believe this inner reality must be vibrating faster than light. F. David Peat is among those who infer no activity as such is involved. From the perspective of the Principle of Correspondence, one would expect such channels to be varied, but comparable to those we know. Though radio waves were always present, now that they have been "discovered," we can use them to transmit sound. Gaining broader vision will likely "reveal" forces linking all beings and species and realms that we now mistakenly label "inert."

Balance in the universe is sometimes obvious and sometimes subtle. Playing the ever-changing boundary between *Selfhood* and *Otherness* is a part of that subtlety. Each being plays a dual "follower/leader" role in the dance between chaos and harmony. In our mutual relationship with the cosmos, we are acted upon as we act. As Raimon Panikkar says, "We leave our traces on things, and the things on us." This principle of reciprocal impact is increasingly more applicable, the higher the level of consciousness involved. Conscious *Selfhood* flourishes only when the individual deliberately and appropriately engages both modes in the flow of interaction with *Otherness*.

The insight of connectedness requires careful reflection during its application to daily events. If our focus is solely on the indivisible aspect of reality, ignoring the polarity of gender and its creative rhythm, we may find ourselves believing that all is preordained from the moment of the Grand Rebirth. However, to understand that one is at any moment an intelligent actor in a cosmic play reveals we have choices about finite (quantum) acts, each with their own short- and long-term cosmic implications. In the selection of one course of action over another to close a quantum gap, we exercise our bit of free will in the cosmic play of co-creation. Our individual selves thus co-design a reality which outlives our personal moment in space-time.

NOTES

1. Lemkow, Anna F. *The Wholeness Principle* (Quest Books. The Theosophical Publishing House: Wheaton, Ill. 1990).
2. Sheldrake, Rupert. *Seven Experiments That Could Change the World* (Riverhead Books: New York, 1995).
3. Panikkar, Raimon. "Some Aspects of a Cosmo-Theandric Spirituality." *Actualitas Omnium Actuum* Verlag Peter Lang (Franffurt am Main) (1989): 342-43.
4. Langley, Noel. *Edgar Cayce on Reincarnation* (Warner Books: New York, 1967), Roberts, Jane. *Adventures in Consciousness* (Prentice-Hall: New York, 1975).
5. Breggin, Peter R. *Toxic Psychiatry: Assault on the Brain with Drugs and Electroshock* (St. Martin's: New York, 1991).
6. Lovelock, James. *Gaia* (W.W. Norton: New York, 1988).
7. Michell, John. *New Light on the Ancient Mystery of Glastonbury* (Gothic Image Publications: Glastonbury, England, 1990). Also see Marciniak, Barbara. *Bringers of the Dawn: Teachings from the Pleiadians* (Bear & Co.: Santa Fe, New Mexico, 1992) and *Earth: Pleiadian Keys to the Living Library* (Bear & Co.: Santa Fe, New Mexico, 1995).
8. The Roper Organization. *Unusual Personal Experiences: An analysis of the data from three national surveys* (Bigelow Holding Corporation: Las Vegas, NV 1991).
9. Bartholomew, Alick (ed). *Crop Circles—Harbingers of Change* (Gateway Books:

Bath, UK, 1991); Dew, Beth (ed). *Gophers in the Crops*. (Gateway Books: Bath, UK, 1992).

10. Von Ward, Paul. "Consciousness Input in Crop Circle Formation." *Proceedings* (International Forum on New Science: Ft. Collins, Colorado: Sept. 1992).
11. Jahn, R.G., and B.J. Dunne. *Margins of Reality: The Role of Consciousness in the Physical World* (Harcourt Brace Jovanovich: New York, 1987).

Chapter 9

Metascience: New Renaissance

Keeping your perspective in mind sustained me in writing this book. By the time you read it, you will likely have had some information or experience that has persuaded you that humans are not the sole conscious beings in the universe. Whether it comes from NASA's Mars study, scientific research, news stories, or personal experience, you will be facing the challenge of redefining your role in the cosmic drama. I hope you find my work supportive. While not alone on the stage, humans do have a particular role: with access to all realms of consciousness, they are charged with co-creation of space-time. Awareness of the larger reality endows humans with the independence to improvise their own destiny.

WITH THE WANING OF THE TWENTIETH CENTURY, the human species is facing a fundamental challenge to the old order that is energizing a New Renaissance with global reach. The old order is being challenged by a revolutionary change in consciousness more liberating of the human spirit and more sweeping in its social implications than the European Renaissance of the fifteenth and sixteenth centuries. The birth of the new cosmic perspective expounded in this book is a greater revolution in human thought than in the Copernican shift in the middle decades of the sixteenth century. In the preceding chapters, this book has presented the elements of the new *metascience* that is replacing the Newtonian and Cartesian assumptions undergirding

Western science, technology, and social systems. These elements offer humans a more creative role in renewing the planetary society and fulfilling their destiny as a cosmic species. This final chapter explores specific ways a metascientific perspective can transform the ways individual human beings think and live and the ways they organize their collective lives.

As we know, institutions and societies can be built on theories or assertions for which there is no evidence of validity. People ignore their own intelligence and inner wisdom when they acquiesce to those who claim divine or official authority. Now that more of the full human story is being rediscovered, humanity must rethink the implications of the biased history which has been written by those with vested interests.

We all have been faced with pretexts of such interests: It is God's will, plan or commandment! The origins of this holy book, and therefore my interpretations of it, are divine! Our family has owned this land from the beginning, or we were given it by the crown, who had it from God! This is the government ordained by God and we are the anointed representatives! Since we are officials we understand the problems and issues better than anyone; our perspective must prevail! This is the truth according to Science, and it has been interpreted by people authorized by the university or government laboratory (chartered by those with "divine" or "official" authority) to speak for Science! And the list of authoritarian assertions goes on and on.

When anyone works up the nerve to raise challenges, they are belittled by those who benefit from the conventional wisdom. And others who fear loss of their own status often join the censure. Examples of this are putdowns appearing in the July 8, 1996, issue of *Newsweek*: philosopher Jean Houston was belittled for innovative use of historical personalities to stimulate the creative thinking of Hillary Rodham Clinton; former NASA scientist Brian O'Leary, for publicly pleading that we take seriously the mounting evidence of UFO activity. Such routine censures and many more serious sanctions mean the established past chokes out alternative futures.

An overarching purpose of this book is to give an expanded foundation for thinking about the individual's access to truth and power. Neither conventional science nor religion recognizes the degree to which humans are dependent on the whole nor to which each individual has the power to influence the whole. With independent access and capacity to validate the knowledge claimed by any other being, each Solarian is equal to all others as a co-creator of life in this galactic neighborhood and in the power of interpretation and understanding of the metascientific synthesis. This truth is the foundation of the Solarian Legacy to be inherited by all who enter the twenty-first century.

Avoiding Catastrophe

The realistic review in Chapter 3 of the natural, and apparently intelligence-influenced, history of the solar system and the Earth is a compelling case for human concern about the integrity of our home base. We are living in a volatile vehicle spinning through the heavens filled with dangerous debris. Two areas call for attention: the potential impact on Earth of an off-planet event (comet passage, meteor collision, solar flares, radiation bursts, etc.) and Earth changes (polar shift, geologic activity, or climatic transformation).

Taking the second first, when the impact of human activities on places like the Middle East and North Africa (where deserts have replaced verdant forests) and the atmosphere (with the greenhouse effect from our fossil fuel energy emissions) is considered, we see clearly our contributions to the demise of our own habitat. Given the indications that previous civilizations, on Earth and Mars, may have also wreaked widespread destruction, there is sufficient cause to be very careful when introducing new technologies. Governments appear to have sufficiently frightened themselves with the potential for a nuclear Armageddon, but societies have not yet adequately focused on the damage overuse of industrial technology is causing to the planet's resource base and ecosystems.

Since Plato, Western scholars have been intrigued by reports of the violent end to the continent and civilization known as Atlantis. We need an understanding of the endings of civilizations in Machu Picchu and related Andean cities, Easter Island, Pyramidal Egypt, and Atlantis. All cultures have residual memories of the flood that almost wiped out humanity. Study of the demise of predeluvial cultures in various parts of the world could help ascertain whether natural forces or the activities of intelligent beings precipitated their destruction.

American and Russian experiments with weather modification through electromagnetic devices, and more recently the defense system use of very low and extremely low frequencies (VLF and ELF) for "Star Wars" programs (communications, radar, and electronic countermeasures), have raised serious concerns about the state of the Earth's ionosphere as it affects human health and the environment. Before proceeding with such technologically enhanced projects, scientists should investigate the causes of the destruction of ancient cities in what is now Iraq, Turkey, Jordan, Lebanon, and Israel. The possibility that atomic or comparable warfare over parts of the Middle East and India (described in Jewish, Greek, and Hindu legends) resulted in areas barren of and inhospitable to life is one that should be studied for the lessons it could teach today's policy makers.

A primary objective of future Mars missions should be to test the hypothesis of destructive behavior, conscious as well as unintentional, having contributed to current conditions on that planet. Mars may give insights into the mistakes of earlier civilizations and ways to practice better stewardship of Earth. If, by terraforming, humans can reverse some of the processes that made Mars so desolate, we can learn how to avoid similar disasters on this planet.

In addition to taking their own behaviors seriously, people must also maintain a watch on the heavens and prepare for the possibility of having to cope with a direct hit or near miss from elsewhere in the universe. The first step is simply to recognize possible vulnerabilities and accept responsibility for avoiding the extremes of social chaos that would result from an unanticipated calamity. The second step is the sharing of resources and knowledge among institutions and countries to insure humanity as a whole is as well prepared as possible. The European Community sky watch program and limited U.S. efforts, monitoring the movement of bodies that pose a potential danger for Earth, are good ideas. Next comes contingency planning, in order to maximize the possibility of an effective response. This does not mean humans should live in constant dread, but all should be aware of the realities of living in a stellar neighborhood where risk exists.

Rediscovering Ancestry

The New Renaissance, surpassing the European Renaissance's rediscovery of classical civilization, will uncover much of the planet's prehistory.

For decades now the evidence has been mounting that human beings are not the sole conscious inhabitants of the universe. But the widely publicized hypothesis that one-celled life has existed on Mars, as simple and as tentative as it is, was the long-awaited signal that it is legitimate to entertain the idea of a broader reality. Although coverage by the mainstream media of the discovery gives the impression that if life existed on Mars it must have been very primitive, people will soon begin to realize that enough time lapsed on Mars to permit the development of life forms not unlike those the Earth has experienced. Once those assumptions begin to circulate in the *noumena*, the possibility of even more radical concepts, like some described in this book, will no longer be considered so farfetched.

In less than a decade we could have concrete evidence of past intelligent life on Mars, if not contact with extant beings. The United States launched two probes in late 1996, to reach the red planet in the second half of 1997. The first to arrive, Pathfinder, has sent back to Earth the first photos from that planet since the Viking missions of 1976. Japan plans a launch in 1998, with additional missions, including landers, scheduled by the U.S. in 1998, 2001, 2003, and 2005. A human expedition is possible as early as 2012. The "Rudolf Steiner scenario," that Mars was one living habitat destroyed in the Solarian odyssey of humanity's ancestors, may be validated as human astronauts discover evidence relevant to their own history. Such Martian discoveries will turn the focus back on this planet and the manner in which people are destructive of themselves and the planet's ecosystems. The history of our ancestors there, perhaps calling for atonement, may show us the need to nurture Earth more carefully.

Discoveries of ancient cities and monuments on Mars, or artifacts on the moon, would also stimulate serious study of similar sites and legends about them on Earth. As described in earlier chapters, ruins of ancient cities, oral and written legends, technologically advanced prehistoric artifacts, geologic records, information received from the Akashic and other noumenal records, and subtle communications with other species of conscious beings all point to ancestors of the human race who lived and were active on Earth millions of years ago. The evidence, while significant, is not yet available in any coherent form; it is therefore impossible to be certain whether humans are natives or the offspring of colonizers. Nevertheless, given the discoveries at our fingertips, all too briefly summarized in this book, it is imperative that we revise humanity's existential myth to provide for prehistoric, intelligent planetary, stellar, and galactic parentage.

Sitchin and Cabrera and others have started the research, but there is so much more to be learned. Sitchin has revealed a center of joint ET/Earthling activity in the Middle East that answers many questions about the stream of advanced life that eventually led to so-called Western Civilization, but his work does not account for other centers in Asia and South America. Cabrera has tantalized us with evidence of even more ancient joint cultures, but his research has not yet garnered the widespread interest and support it deserves. Researchers into the mists of human history need a large, multidisciplined, multisense effort to fill the gaps in their analyses and to incorporate evidence from all parts of the globe.

One of the most crucial mythic reexaminations called for is that of the current doctrines of governance. Sitchin makes a strong case in his series of books that the visiting "gods" instructed humans in concepts for social institutions, in addition to hard science and technologies for practical living. These institutions included the concept of allegiance by groups of humans to aliens, cum royal beings, who commanded their support in internecine alien conflicts. (According to Cabrera, adoptions of similar alien tactics may have led to tribal divisions among the pre-Incas of Peru.) In this context, the subsequent appointments of local human leaders by the "royal" beings would certainly have given rise to the idea of kingship by divine anointment, and, by implication, warring tribes. (It is feasible that such a single divisive concept could have caused the human race untold suffering over the last two thousand years.)

Even modern concepts of democracy, including "government by the people," have been distorted by historical bias in their implementation. These innovations were grafted onto old structures derived from a notion of "divine officialdom," a mindset that justifies the keeping of secrets and resources from public knowledge and holding back alternative developments. How many more current institutional defects are the result of such exogenous interventions? Were patronizing colonizers responsible for giving humans a sense of being flawed, of being abandoned orphans, inclined to intrafamily squabbling, and addicted to dysfunctional rituals of adulation for absent parents, sometimes called gods?

For example, if our ancestors were abused by aliens, modern humans need to come to grips with the physical and psychological imprint of having been treated like inferior aboriginals. What scars have such experiences left on our collective psyche? How can humans assuage the guilt of selling their sisters to higher species and groveling as their servants? Perhaps the impact has been more destructive to human than intraspecies enslavement, even to the extent that interspecies patterns led to such enslavement among ourselves.

Conversely, what are the positive human attributes that can be found when mentally regressing into our prehistoric experience? Are we the repository of values and insights that can enrich life in the Solar neighborhood and among our galactic cousins beyond? What are our unique offerings to the ongoing spiral of cosmic life? Are we to become the sentient and sapient race to maintain this planet, reseed Mars, and export the sparks of organic life and cosmic consciousness to barren planets afar?

If our forebears were the manufacturers of the sophisticated artifacts recently discovered from antiquity, why does much of the technology differ from our current levels and forms? Was it based on entirely different principles, outside modern science's current understanding of natural laws? The capability to move structural and monumental stones weighing many tons (found on numerous prehistoric sites) without apparent mechanical power is one example of the application of unknown principles. Knowledge about the impact of the Earth's magnetic field and its extraterrestrial gravitational forces on the lives of Earthlings is another. The related ancient discipline of astrology calls out for new scientific research. Ancient flying machines reportedly using the natural forces of gravity and electromagnetic fields are outside current technological capabilities and need to be publicly researched. In instances like these, a mother lode of promising knowledge in extant texts and artifacts from earlier civilizations is waiting to be mined.

Ancient wisdom can prove helpful beyond the development of machines and energy sources for transport and mechanical power. Society can even learn from the Sumerian tablets' codes for behavior focused on justice, instead of the later ones of Hammurabi focused on crime and punishment. With serious concerns about many aspects of the American system of criminal justice, society could benefit from a second look at ancient wisdom. Medicine is another rich area for such a process of rediscovery. Much has been preserved through the folk traditions of many societies. Instead of simply dismissing out of hand the claims of medicine men and women, or totally accepting their views, the healers of the future will use modern concepts and technology to evaluate folk medicine and incorporate into their practice whatever therapies work. Combining perennial wisdom, predating the current Nicene-Newtonian era, with new science insights could revolutionize our understanding of human psychology, including our ability to conceive and manifest changes in the phenomenal realm.

Implications of Metascience

Metascience requires that humans use all senses (physical and subtle) and all sources of wisdom. To exclude *a priori* any body of knowledge, or to fail to take advantage of any sense, dooms humanity to a diminished life, less than the full birthright of the Solarian Legacy. Three examples will illustrate the value of *metascience* combining ancient wisdom with frontier science: (1) Applying the notions of ancient pantheism and a singular consciousness to teaching people how best to relate to the environment. (2) Enhancing practical applications of concepts like telepathy and intuition through conscious development of the subtle senses. (3) Explaining the uses of traditional practices like I Ching and divination in terms of biocommunications and synchronicity. Further implications of *metascience* follow.

In the macrocosm, the Principle of Cause and Effect controls the play between the Moon and the tides, between the stars and our solar system. In the microcosm, it directs the creative dance between mind and matter and between the interaction of time and the dispersion of energy. As the physical universe follows a cycle of birth, growth, decline, death and rebirth, synchronized by the Principle of Correspondence, so do individuals, societies, and interspecies systems. As matter and antimatter are balanced throughout the universe,* so is the yin and yang of human behavior. Greater understanding of these dynamics will enhance our freedom from and use of the material systems at the same time it increases our respect for our place in them.

The Principle of Polarity functions in the positive and negative charges of nuclei. It is seen in the contrasts of centripetal and centrifugal spins (cyclones and hurricanes), or the rise and fall of acid/base ratios. Love and anger, good and evil, are polarities, not unlike the inseparable twins of particles and antiparticles. We may have misinterpreted certain emotions and therefore misguided ourselves by not recognizing that they are simply poles of the same behavioral force.

The science of biotechnology—like the phenomena-based field of physics—is circling back to the nonphysical underpinnings of matter explained by the three-faceted model. Just a short decade ago, the young industry used genetic engineering to create new drugs to complement the body's systems. Now, some biotechnicians, working cooperatively with large drug manufacturers, are using computer modeling to design compounds, molecule by molecule, that hook into isolated diseased cells and modify their behavior. Soon these scientists will begin to see evidence of their own conscious interaction with the cells under their scrutiny. When they do, the circle will be complete. Intuitive delving into the material realm will have led them back to the power of their own minds. The ancient Ayruvedic concept—that the expression of intent can create physical changes—will be validated in gleaming metallic research laboratories.

Legend has it that ancient Hawaiian goddesses were shapeshifters—able to change form at will—as were aboriginal shamans. Their expressions of intent from the realm of consciousness gave rise to patterns of force *(energeia)* that worked their will on *matenergy*. These and other miracles of living men and women can be routinely studied alongside anomalous aspects of the UFO phenomenon, the manifestation and disappearance of

---

* Scientists have not yet perceived this balance, but some groups, such as the astronomers at Arizona's Whipple Observatory, are measuring antimatter through various surrogates.

matter, and the bilocation of people. Searching along the arrow of time will become plausible as precognition and recall of past events are studied on presently unknown spectra parallel to the travel of subparticles and the flow of ordinary energy. Evolving cosmic humans will join other beings already able to flit from realm to realm in subtle form.

Using All Senses

In the integral but three-faceted universe, it is possible for humans to see the heavens through electronic telescopes or from spaceships *and* also confirm their reality through the subtle senses, without the aid of technology or leaving home. Success in subtle sense exploration of the workings of the universe will reverse the modern human tendency to let these senses atrophy through excessive focus on the rational and physical. As they become receptive to more cues, humans will open themselves to the gifts of synchronicity, recognizing its power as the multifaceted and biocommunication-based "grantor-of-wishes" or "manifestor-of-intentions." With more facets of *beingness* at their disposal, humans will more readily master new levels of creativity.

In a cosmos that integrates matenergy, *subtle energy, and mind, precipitous leaps to action or reaction in attempts to achieve a desired end will be recognized for their futility.* Behavior incongruent with inner wisdom precipitates reverberations with unforeseen consequences, with the possible effect of undermining the desired outcome. For example, the "legal" killing of people who kill ignores the inner implications of such acts for the whole society. That way of dealing with dysfunctional behavior reinforces the causes—psychological or energetic alienation and absence of conscious respect. Regardless of how many killers are officially killed, unless society eliminates its support for violent acts at all levels (by people in any position of economic or personal authority over others), high murder rates will remain.

Recognizing that any situation is comprised of a set of consequences accumulated from a plethora of prior events and their indirect, inner connections, only a comparable confluence of complex and synchronistic forces can bring about the desired change. Making good social policy is like riding the rapids of a swollen stream to reach a rock on the opposite shore: mastery of society's river is accomplished by making integrated sets of small adjustments suggested by all senses in a given instant. All post-modern institutions could benefit from the leadership experience of metaphysical white-water rafting guides.

Understanding the three-faceted universe gives new insight into human motivation and behavior. An example is the fact that there is an energetic aspect to the need for learning. A vacuum of knowledge or a perceived deficit of understanding (one pole) energizes the impulse to acquire new information or answers (the opposite pole). During the phase of active seeking one is attempting to reach some threshold, a point of conviction, where the kinetic energy is transformed into an "I see" experience. Energetic satisfaction has no qualitative judgment to it: any answers from accepted messengers will do. Answers are the energetic polarity to the quest; they are strongly held until one feels a significant incongruency, and a new round of seeking begins.

All sentient beings, therefore, have an energized resistance to external pressure to shift a mindset. Just as there are ingrained *matenergy* patterns (as in genes, muscular memories, and satellite orbits), embedded emotional or subtle energy patterns, regardless of their validity or current relevance, stay in place until reconfigured by deliberate acts that transmute their polarities.

Successful traditional societies developed individual and group practices that facilitated transmutation of dysfunctional energy blocks. Included were group festivals, ritual dance, or other controlled physical activities. In some societies, to prepare them for self-sufficiency, individuals were sent alone into the wilderness or on a quest to discover new answers for themselves. The more traumatic the instability, the more motivated people were in the search for a new personal order. Consciousness state-shifting rituals and emotionally-laden initiations had similar results.

To think one can simply order the transformation of pent-up energy charges or smother their expression is no longer credible. Exclusive reliance on the external use of force to bring about a shift must be eliminated: whether laws, police regulation, or religious prescriptions, they all harden the resistance or divert it into a different channel of expression. Only when individuals freely participate in their own transformation will constructive transmutation of the undesired polarity (fear, anger, etc.) occur. As forest experts are beginning to grasp the wisdom of permitting small fires that are necessary to the forest's health, thereby avoiding the infrequent, but huge conflagrations that destroy the entire forest, society's future leaders will learn how to facilitate the transformation of destructive energy polarities before they get out of hand.

Mastery of the subtle senses, confirming prescientific ways of knowing, will revolutionize human understanding and behaviors on every level. Such mastery will dramatically change the character of human intercourse, eliminating much of the hypocrisy and deceit prevalent in business and politics, as well as personal relationships. In a turn to fully conscious leadership practices, societal, political, and organizational processes would be very different from our current legalistic, confrontational, and top-down approaches. Until the subtle senses are publicly incorporated into the deliberations of the U.S. Congress and other governmental councils, these bodies cannot effectively serve the humans touched by their actions. How would such governance feel? What would it be like to begin official deliberations by focusing on everyone's inner communications, and responding to them in a cooperative way?

With widespread use of the power conveyed by the subtle senses, officials would find themselves subject to the will of their equally powerful cosmic followers, whose potency resides in their ability to give or withhold their subtle energy support. The inner power of the individual rests in the personal control one has over the opening of energeial and noumenal membranes. Without that mental and subtle energy support from the group, no leader can maintain authority very long.

How different the court system would be if the subtle senses were used to reorient the current criminal justice approach of legalistic confrontation about physical evidence. Imagine how the behavior of defendants and plaintiffs alike would change if the purposes of the investigation and trial processes were to ascertain the truth of the inner and outer realms, rather than what can be proved by the narrow rules of evidence. What if the jury's responsibility was to assign appropriate responsibility for corrective action and compensation to all in the society who had contributed to the problem. If the focus were on personal intent and impact, not on bureaucratic procedures, everyone would become more concerned about the effect of their actions on the whole.

Depersonalization and the monstrous scale of current financial, business, government, and even nonprofit organizations have undermined the self-disciplining forces of community ethos and personal accountability. Formal social control mechanisms such as elections, audits, financial reports, etc., cannot keep up with this erosion of social integrity. What is needed are social inventions that incorporate knowledge of the three realms and intervene in the processes of multilevel co-creation.

One such social invention would be the systematic introduction of the use of the subtle senses into the leadership and management of organizations. Nonintrusive soundings of an organization by skilled and responsible use of subtle communication channels can tap into negative, confused, or blocked energies that presage problematic performance. Intuitive professionals can diagnose the noumenal origins of the energy patterns and relate them to behavioral trends. Trusting in the full range of senses, they can use multiple insights to develop interventions that lead to greater organizational clarity and interpersonal harmony. With the professions of management consulting and organizational development radically reformed, people would have nothing to fear but their own intentions.

The realization of this three-level model of conscious behavior requires an approach the opposite of the Western system of education, in which the objective is to create people who are alike and serve the status quo. Instead of the right answers being given to the students by the teachers, the process should assist the child in realizing the ten-plus sense of his/her own power. The roles of shaman and apprentice are still useful models for the process of cosmic education. The teachers would encourage confidence in the inner wisdom of each being and its innate impulse to wholeness.

Conscious role playing is the best approach to learning this new way of being. As in improvisational theater, the players make up the script in concert with each other. All that is required is a commitment to pay attention to every possible cue, regardless of the sense through which it comes. By knowing that positive energy flows easily through the autonomous nerve/muscle system while the negative sets up resistance, one can easily shift polarities using a conscious reinterpretation of the idea or event. The need for catharsis, the recognition or expression of repressed fears or other negative feelings, on the battlefield and in political or bureaucratic warfare can be satisfied by improvised games or encounters created explicitly for the temporary assumption of different modes of expression.

Currently, certain energy-shifting training techniques used by body workcrs, thcrapists, and drill sergeants unwittingly involve such subtle pattern reprogramming. The challenge for twenty-first century humans is to create collective experiences of consciously designed catalytic events that transform society's inertial energy into a force for openness and experimentation. In other words, people must consciously stimulate and involve the *energeia* and the *noumena*, as well as the *phenomena*, in efforts to develop human potential. As more people learn these transformative methods, a quality of social cohesion will evolve that makes possible planetary transformation, the flowering of humanity's Solarian Legacy.

Testing Limits

The birthright and responsibility explicit in the Solarian Legacy have much broader scope than we have imagined. Attempts to describe with certainty a particular limit in our reality, to logically and empirically prove that any such finite boundary exists, founder for lack of proof. The physical boundaries of cells disappear under the microscope, atoms change their nature, subatomic particles flash in and out of existence, and energy is easily transmuted from one form to another. There are no ideas or memories that one can call exclusively one's own. The boundaries between life and death vanish in the face of other dimensions. Even gravity and time prove flexible in the experiments of frontier scientists. Recent developments described in earlier chapters indicate that breakthroughs in knowledge even more profound than those of twentieth-century physics and chemistry are imminent.

The power of conscious healing will be one of the first breaches in the limits stoutly defended by the traditional scientific paradigm. Many scientists who have felt the power of meditation and friends' prayers, or the energy of psychic healers, are leading the way. Acceptance that invisible thought or energy patterns can heal diseased cells without physical contact will transform the field of medicine. Homeopaths' use of computers to focus and transmit patterns of energy through light waves that potentize (or patternize) clear water will help many to accept the same focusing of healing patterns (intent) by a conscious healer. Individuals in turn will be encouraged to assume such power in their heads and hearts, issuing guidance to their own cells. Those who would be facilitators of healing will soon learn their most useful role is that of reinforcing thoughts of health and the flow of subtle energies that a person permits into his or her body.

Gone will be the days of antibiotics, vaccinations, super drugs, and massive doses of radiation and chemicals that in the long run destroy more than they heal. The physician of tomorrow will return to the realm of *metascience* to incorporate thoughts and emotions into treatments of the body. She (all true healers regardless of gender practice the feminine art of receptive channeling of subtle energies) will encourage the individual to tune into the thought forms and the emotional state relevant to the area of dis-ease.

Conscious interaction with the environment will be another rupture in the traditional mindset. People will master the channels of communication with plants and animals, and establish cooperative relationships with ecological systems. Food scarcity will be a thing of the past as humans consciously cooperate with other species who share their nurturing life forces with them. People will no longer eat matter that has had the life frightened, squeezed, or burned out of it. With multirealm ecological thinking, we will start by envisioning the interactions of our thoughts, feelings, and actions with all beings. Individuals will take into account the perspective of the rivers, the birds and the bees, the earthworms and the viruses, the rocks and the trees, and on and on, when we intervene in their lives. Using subtle senses to augment physical ones, humans will "listen" to the desires of other life forms as well as our own.

In the mechanical realm, there will be general access to the power of manifestation. Consciously directed interaction of the subtle energies with *matenergy* will fuel power transformers and transport vehicles, and many other forms of consciousness-assisted technology. The result will be a fundamental shift in economic systems as subtle energy is demonstrated to be a limitless resource easily available to everyone.

To break the limits humans have imposed on themselves only takes an act of cognition: to perceive that everyone is still as inextricably and fundamentally connected to the universe as the traditional hunters, who believed that in seeking their prey in the wilderness they were also being sought by it. Modern society has delegated to impersonal systems the acquisition of food, the reading of the weather signs, and the provision of mechanical communication channels. While such systems will be difficult to dismantle in urban and suburban centers, control of them will become more community-based as people engage in the maximum use of subtle communications to assert their inner power. By inserting themselves back into the natural flow of group events, people will experience the fulfillment that comes from being a conscious participant. They will feel less alienated from others and more connected with their essential self.

Anyone who desires can learn from examples like that of the Arhuaco Indians who live in the Sierras Nevadas of Colombia, South America. They continue to hold a world view that predates the arrival of Columbus: living correctly is being in harmony with the natural principles of the universe. They believe illness or other problems occur when the laws of nature are violated. Their earth is alive, filled with an inner spirit that transcends time and finite matter. Everything, including rocks and all beings, lives forever, with only shifts in form and place. The right frame of mind is crucial in starting any action or interaction: if one begins with the wrong intention, nothing goes right.

When we recognize that basic assumptions or intentions shape the elements of life, one of the conceptual barriers that must be examined is the artificial division we have created between the institutions of spirit and of governance. We should not invite dogmatic religions, as they currently manifest themselves, into public positions of authority, for all the reasons previously discussed. However, the American relegation of the realm of the heart and the other subtle senses to the purview of the church, restricting government to the five-sense, left-brain, mechanistic mode of thinking and problem solving, has deprived politics and other public discussions of constructive input from the energeial and noumenal realms. Citizens need to find a way to introduce *browing*, *hearting*, *splaning*, *shading*, and *rooting* (see Chapter 4), as well as the senses of consciousness, into the process of public dialogue and political discourse.

In the environmental arena, where there is apparently no *a priori* limit to our destructiveness, the only constraints on human behavior toward the planet depend on the exercise of personal judgment. Given that the current state of the Earth is the result of countless generations of intervention by conscious beings, perhaps even to the point of creating myriad species, there is no question of letting the planet go back to a natural state. As apparent stewards of the planet we must devise satisfactory alternatives to the destructive trends and seek the support and concurrence of other species. Only with their cooperation are we likely to succeed.

Cosmic Fraternity

Humans become qualified for membership in the guild of cosmic beings as they gain multilevel awareness of the vast, conscious, 20-billion-year-old universe into which they have incarnated. The price of admission is a simple declaration of independence from any would-be rulers who wish to deny, or are blind to, the Solarian Legacy. For unknown reasons, perhaps our earlier immaturity, older races have been partially responsible for keeping humans in the dark about their true nature. However, part of that myopia is self-imposed: there is no longer any excuse for failing to outgrow adolescent innocence and assume our cosmic inheritance.

Adults cannot escape their cosmic responsibility by blaming governments who fail to share the knowledge they have and refuse to engage in frank conversations with citizens: official secrecy does not prevent efforts to correct the human story. If covert human organizations are actively involved in *sub rosa*, and perhaps cooperative, projects involving aliens, the facts can be publicized and brought into open discussion. If technology transfer is underway in hidden facilities where military forces, using crafts recovered from UFO crashes or interspecies technology transfers, are gaining access to advanced material, design, and propulsion system concepts, there is no reason not to publish the news for all humanity. The fact that government officials, from the president on down, refuse to engage in a dialogue with responsible and well-informed citizens on the UFO/ET issue only tosses the ball back into the private sector. Private individuals and groups from various areas of American society can take the initiative to bring about more honesty and openness.

Even if all the answers are not yet clear, the limited information now widely available to the public is enough to call into question official motivations and attempts to suppress information. We are at a point where a new covenant, providing for joint validation of important truths, must be negotiated between the electorate and those who would assume positions of leadership. Citizens can forgive deceptions in the past and forge a new bond of cooperation with institutions that pledge to obey the new rules—universal access to all knowledge and respect for the integrity and interdependence of all beings. (The recent U.S. Air Force release of a new Roswell crash-site report, claiming that people reporting having seen alien bodies must have seen crash dummies, fails to meet either of these criteria.)

Like African teenagers approaching Poro initiations and rites of passage to adulthood, we may be anxious about giving up comforting delusions, but humans are much more resilient than we are given credit for. People will welcome being part of a historic opportunity like the dawning of an unprecedented age of understanding and the advent of the universal family. The predictions of mass fear are no longer well founded: there will be no repeat of the widespread reactions to the 1938 provocative radio broadcast of Orson Wells' *War of Worlds.* There is little evidence during the last half-century that the potential for social hysteria and chaos is disturbing enough to justify further secrecy or hesitancy in unveiling the Solarian Legacy.

Informal surveys show that many religious leaders believe they can incorporate knowledge of ETs into their theistic concepts. But their followers are way ahead of them in understanding and accepting the vision of an even larger order. This is demonstrated by widespread interest in movies, TV shows, books, newspaper articles, and conferences portraying alien involvement in human affairs, varying from hostile (*Independence Day*) to benign (*Michael* and *Touched by an Angel*) to satirical (*Men in Black*). This deluge of information, combined with the so-called extraordinary experiences that all share, has made it easier for everyone to entertain new ideas about the nature of reality. Most people are now primed to progress quickly beyond the age of cosmic adolescence. They are ready for the excitement of working out new forms of interaction with partners from all planes of existence.

Isaac Asimov invented a multispecies universe to serve as the background for multi-authored science fiction stories. He posited the emergence of six sentient races, from among hundreds in the galaxy, with the capability of interstellar flight. Among them were the newcomers Erthumos (ourselves), "the warm-blooded bipeds from the third planet of an obscure, third rate sun." These six races represented widely different cultures and technologies, springing from beings whose physiologies were as distinct from each other as human imagination could make them. In stories set a thousand years in the future, these races manage to abide by an accord of galactic peace, and serve as models or development specialists for less advanced, planet-bound peoples.

Humans actually have the potential to fulfill that visionary role: we must begin consciously rehearsing for the ultimate family reunion (as in the 1995 conference in Washington, D.C., on the knowledge and skills needed for meeting alien cultures, sponsored by the Foundation on Human Potential). People have demonstrated how quickly they can reorient their thinking about former enemies (Americans, Germans, Japanese, Russians, and Vietnamese), to actually embrace them as friends and new business partners. With a capacity for such transformations, humans may be good candidates for galactic leadership roles in a fraternity of cosmic cultures.

As humans come to recognize their Solarian Legacy and take cognizance of nonhuman cousins, what are the psychological and behavioral implications of this rite of passage? Self-consciously admitting themselves to the fellowship of a more mature cosmic community dramatically broadens perceptions of roles and responsibilities, and thereby enhances the joy and pleasure from daily life in this incarnation. *The expansion of thinking enhances not only the sense of doing, but the power of doing!* Plus, individual strength is multiplied by orders of magnitude when large numbers concentrate their energeial and noumenal forces for the same purpose.

Moving from a state of Earth-bound somnambulism to an appreciation of a fully living universe activates underused inner powers. Knowing they possess such powers, humans cannot help but behave differently. The old approaches of seeking personal advantage, using any manipulation necessary, and attempting unilateral control over events are recognized for their ineffectiveness in the multileveled universe. In such a reality, impulses to covertly manage processes and people in one's own interest, or even in service to family and community, are seen as intellectually and ethically immature.

The interactive process is simple and easy to understand. The perpetual need to give and support, the expressive pole, has an integral counterpart in *Selfhood*—greed that seeks self-fulfillment in the dark of the night when inner defenses are down. The latter may lead us to believe we have earned our gains with no need to share, but we cannot escape the balancing forces of the three realms. That which is gained in one form must also be given in another. Karmic justice comes calling when we halt the open give-and-take and try to close an account in our personal favor.

Understanding the connections that bind all beings together, one learns to interpret the waves of subtle energy from other beings. Recognizing that they too wish to maintain their own integrity, mutual respect for jointly designed outcomes replaces attempts to manipulate *Otherness* for selfish purposes. Respect for physical, emotional, and mental boundaries results in the reactions of others becoming more positive and supportive, with everyone requiring less expenditure of physical and subtle energies. In these circumstances, decisions and actions based in the multileveled knowingness of the subtle senses are found to be more satisfying. Such positive and powerful interactions make an immediate difference in daily life: people accomplish more with less, and feel better doing it. Instead of lonely self-promotion as unique stars, all cosmic actors can glory in the camaraderie of joint success.

Becoming aware of one's co-creative power gives rise to a sense of responsibility for its use. Knowing the capability to unleash subtle energies that wreak havoc on unprotected others also has the potential to destroy oneself, conscious beings learn to exercise judicious constraint. The functioning of the Principle of Cause and Effect makes it impossible to escape the ramifications of one's intentions: for beings with the inborn power humans possess, the universe guarantees accountability by individuals, groups, and nations. Conscious humans have no choice but to cooperate in the interest of their own long-term well-being.

A leisurely pace for developing interplanetary cooperation may be fine in Asimov's fiction, but Earthlings do not have another thousand years to prepare for membership in the cosmic fraternity. From within human society come the challenges of interracial, intercultural, and intragroup breakdowns that are straining social institutions on every continent. At any moment, humans could be faced with learning to interact constructively with a variety of sentient beings: co-habitants of Earth, co-Solarians, or beings from without this solar system. Humans have no choice about moving into the cosmic arena: the internal and external pressures to expand our vision cannot be held back.

Current society's twentieth century, an explosive period of emotion-charged creativity, must certainly have been or be of interest to older, more advanced cultures. They may seek overt dialogue if we do not seek them, providing answers to questions we did not know we had asked. Perhaps we are seen by other civilizations as the "generation to come," with ideas and energy that will change the course of Solarian or galactic development. We could be offering them as much hope for the future of the larger community as our children do for us.

Membership Conditions

Knowledge of their Solarian Legacy and their inner powers makes it easier for humans to welcome the curtain's rising on this new intergalactic era of cosmic creativity. But there is still uncertainty about the conditions of membership in the cosmic fraternity. Even the tentative incorporation of the possibility of nonhuman actors into one's thinking forever changes the perception of roles. Humans may be concerned about being upstaged or replaced in their perception of a primary role for themselves on the planet and in the solar system. They do not yet know the power and attitudes of their fellow cosmic beings. Whether their intent will be hostile, cooperative, supportive, or disdainful of Earthlings' level of development remains to be seen. With no clear idea of what to expect in terms of languages and appearances, their personal habits, or their social norms, how can humans prepare for all the possible scenarios?

Metascience in a three-level universe provides some basis for reasonable assumptions. First, it appears that obvious differences are in form, not substance. The term "alien" is inappropriate for beings who are fundamentally like us—in reality our cosmic siblings. All incarnated beings, regardless of planet of origin, are irrevocably linked to each other beyond space-time at the emotional and mental levels.

One likely surprise for many people will be the discovery of the parallels between humans and other cosmic beings in the area of psychic capacities. One of the desirable, and probable, outcomes of recognition of our cosmic siblings, in spite of surface differences, will be a change for the better in attitudes and behaviors among members of the human family. We will recognize, like the early Mahayanist reform movement in Buddhism, that in the search for greater illumination and wisdom, we must consciously improve the performances of all before any individual can fully progress. Most likely, more advanced beings will have already learned this lesson, so we should begin to put it into practice among humans before we get on the larger stage.

The physical differences among cosmic siblings with which humans will have to cope will generally require only adjustments in thinking. While this incarnation may be one of differences, humans may recognize beings they have known in other incarnations. This recognition is possible because the physical life spans of beings from other planets are likely to be a function of some factor such as the duration of their home's orbit around its sun, but the longevity of noumenal bodies are probably comparable across the galaxies.

One of the confusions that apparently arose in earlier human contacts with extraterrestrials was the impression that they were immortal: to ancient humans, the normal life spans of beings whose home base years were so much longer than humanity seemed like eternity. Judging from the perceptions of human civilizations during the period from 4,000-1,000 B.C.E., the visiting beings chose not to share knowledge of differing life spans with humanity. Some humans came to believe that the offering of extended life to a few faithful human servants was the gift of immortality. That understanding has been used to buttress belief in misleading religions by holding out the promise of eternal life to those who die in service of their causes. Loss of such a lure should return life's focus to the cosmic here-and-now.

Another result of humans gaining a sense of their full membership in the cosmic community will be the demise of the other fantasies created by the leaders of religions that are followed by millions of adherents. All the debate about the appearances of various "gods" and their miraculous feats, explained by Zecharia Sitchin and historians like him, will become moot. When people understand that tales in religious texts of seemingly divine travel in and communications from the air, and other seemingly miraculous feats, were only the normal activities of advanced civilizations, there will be no support for belief in divine commandments and special dispensations from such ordinary cosmic beings. When the concept of Yahweh is seen as a composite representation of several senior officials among visiting interstellar travelers, his alleged rules, and those of Mohammed (perhaps another such traveler) and other religious leaders, will no longer be taken at face value. Only scriptures that have experiential relevance to our multi-dimensional lives in this space-time will be considered worthy of teaching our children.

This means that religions still viable in the twenty-first century, those truly re-linking humans to their inner realities, will divest themselves of "divinely inspired" scriptures and holy books, eliminate their historically rooted hierarchies, do away with "special revelations," cease artificial ritual, and profess no hidden power. Instead, they will form communities of seekers after truth, using the minds and experiences of all members to allow for the broadest possible use of noumenal knowledge integrated with daily experience. They will invite other communities of seekers to share their own findings and test conclusions of any group in public forums.

Conscious Living

As one gains a fuller appreciation of the intricate and multidimensional nature of cosmic consciousness, the vision of performing at a higher level, consistent with this new knowledge, becomes a motivating force. To act routinely when there has been no opportunity for or awareness of greater possibilities is excusable, but not to attempt one's potential after the dawn of self-awareness leaves one feeling unsatisfied. For example, now that we know about multiple-leveled senses, to live without taking full advantage of them is analogous to engaging in a boxing match with one hand tied or trying to paint by numbers without the color code. We feel the frustration of functioning with only half the resources available to us, and that knowledge incongruency (as discussed earlier) feeds the quest for new information and experience.

The quality that distinguishes the conscious performer from the uninitiated one is a desire to continually expand the circles of awareness of both the inner and outer worlds, using all the senses or gateways between the realms to enhance the richness of the incarnation. The ability to shift frequently, maintaining an overall balanced state, lessens the need for drugged sleep or artificial mood changers and converts the pace of waking life to one of sustained alertness to subtle information. This enhanced state of consciousness provides more options for deliberate self-direction in the management of one's own life, in collaboration with other conscious beings.

Humans living in full consciousness of their true legacy and its potential are awakened to assume full responsibility for the quality of their existence. Life is seen as a cosmic dance, in which certain capacities and patterns of behavior are intrinsic to the incarnation, but where the design of the specific steps is left to choices made in conjunction with one's partners. *Selfhood* uses all the senses, stretching to the fullness of all polarities: anguish and joy, satisfaction and guilt, initiating and reacting, holding steady and abruptly changing course, and receiving and expressing. A single incarnation has many stages with all these qualities, with plenty of time to learn and practice. In maturity, the being, positioned and energized by the experience of almost a lifetime, clearly portrays the sum total for all to see, ready for a rebirth in the cosmic cycle.

Living consciously takes only a few crucial assumptions that provide the basis for self-empowerment. They involve appreciation of the power of conscious thought in shaping all aspects of life:

- Any person has all cosmic powers that are inherent in others
- Each being can test the assertion of knowledge by any other
- Many developmental paths lead to self-realization
- Conscious choice is required to stay on a chosen path
- Intentions are manifest through both subtle and physical senses.

One implication of these assumptions is that the wisdom one has about one's self deserves the total respect of others. A corollary of this is that when two or more beings decide to act in concert they have equal access to wisdom, and should equally share in the design of their chosen joint ventures. Individuals should be careful not to define joint ventures in a manner that infringes on their valued inner boundaries. The maintenance of that equality in the interconnectedness leads to the next insight: Social progress comes from learning how to balance between the two, providing for both individuality and cooperation.

In the final analysis, what will it take for humans to step out on the cosmic stage in a significantly bigger role? Self-confidence! In one of life's great paradoxes, the only way to calm our trepidations about taking on such new responsibilities is to give up our long-held bastions of security. Acceptance of the Solarian Legacy provides only the essentials we need to assume the mantle of *cosmic beingness*. The rest is up to us. I offer the following meditation on what putting on the cloak of cosmic being has meant to me, for whatever resonance it may have.

No longer wrapped in the anthropomorphic arms of a father-figure, protective god of my childhood, no longer inhabiting a predictable universe in which my brothers and I had dominion, I, ironically, now feel more at home and secure. Gone is the fear of a vengeful wrath if I fail to maintain a guarded, hesitant approach to life. Gone is the sense of puppet-like, one-way strings that could jerk me off stage as easily as they make me dance. Released from admonitions that I accept on blind faith the answers of a few who claim divine connections, my curiosity quickens as I realize partial answers to any question lead to a more profound, more illuminating plateau of untold possibilities. With the end of my adherence to traditional religion and conventional science comes a metascience *of awesome scope, literally connecting me to the powerhouse of the cosmos.*

My sense of being part of a seamless universe, where anything is connected to everything else and where each being is securely embedded in an eternal and vibrant pulsing of cosmic life, is even stronger now than it was in the cosmology of childhood. While recognizing that my life is more complex and its unknown interweaving more intricate than the explanations of any religion or current intellectual school of thought, I nevertheless am more psychologically at ease than ever before.

I see a marvelous, galactic field of flowering plants and gamboling animals, adorned by the stars and planets in a matrix of patterns and forces that are visible to the mind's eye and link all elements to each other. I see all of us playing both on stage and in the wings, adding our own variations of form, movement, sound, and color. We change roles at will, moving from that of actor to that of cosmic artist (with ourselves simultaneously in the palette of colors and among the images emerging from the canvas). With my inner eye open, I can no longer avoid "seeing" the luminal and subluminal vibrations of all the organic and inorganic entities around me. The boulder beside the path is as vibrant as the tree; its patterns, more stable in a field of flitting forms, helps to ground me to our material planet, and at the same time releases my vision, inner and outer, to soar to the stars. I feel myself suspended among all facets of the cosmic diamond, like the astronaut in microgravity. The winds of subtle forces cause the bodies to waft back and forth throughout the rhythmic patterns of night and day, life and death.

I stand under the starlit sky and glory in the existence of innumerable conscious beings clustered throughout the universe. I am drawn to merge my partial stories of creation and visions of the Grand Couple with my cosmic siblings' perceptions of our origins and our cosmic legacy. I no longer wish to hide in my human uniqueness, but yearn to reach out to the whole conscious family, across the galaxies and through the dimensions, to pool human wisdom in a partnership of beings that vaults everyone to the next expansive plain of cosmic awareness.

I imagine picnicking together around the everflowing lake of the cosmic void filled with potentialities beyond imagination, aware that the manner in which we play the games of the day will determine the health of the cosmic fields of tomorrow. Knowing that the pace of universal growth overshadows the span of a human lifetime, I, nevertheless, sense that each being's incarnation is an essential link in the unbroken chain of the future. I feel the reverberations of our creative and destructive efforts will live on in all realms until future generations are able to transform them.

Why does admitting to so many gaps in knowledge, giving up the safe conventional theories, result in such a sense of release and ease? I believe it comes from dropping the childish pretense of knowing things we do not. Moving beyond pretense gives us a full measure of self-confidence in the value of our Solarian Legacy, and in our collective ability to realize more fully its potential in the twenty-first century.

SUGGESTED ADDITIONAL READINGS

1. Zubrin, Robert. *The Case for Mars: The Plan to Settle the Red Planet and Why We Must* (Free Press: New York, 1996).
2. Narlikar, Jayant. *The Lighter Side of Gravity* (Cambridge University Press: New York, 1996).
3. Reiss, Michael J., and Roger Straughan. *Improving Nature? The Science and Ethics of Genetic Engineering* (Cambridge University Press: New York, 1996).
4. Micozzi, Marc S. (editor). *Fundamentals of Complementary and Alternative Medicine* (Churchill Livingston: Philadelphia, Pennsylvania, 1996).
5. Davies, Paul. *The Mind of God* (Simon & Schuster: New York, 1992).
6. Loye, David. *The Sphinx and the Rainbow: Brain, Mind, and Future* (Shambhala: Boulder, Colorado, 1983).
7. Eccles, John (editor). *Mind and Brain: The Many-Faceted Problem* (Paragon House: New York, 1985).
8. White, John. *The Meeting of Science and Spirit* (Paragon House: New York, 1990).
9. Ornstein, Robert, and Paul Ehrlich. *New World/New Mind: Moving Toward Conscious Evolution* (Doubeday: New York, 1989).
10. Ferris, Timothy. *The Mind's Sky: Human Intelligence in a Cosmic Context* (Bantam Books: New York, 1992).
11. Grosso, Michael. *Frontiers of the Soul* (Quest Books: Wheaton, Illinois, 1992).

INDEX

ABOUT THE AUTHOR

Paul Von Ward's previous book is *Dismantling the Pyramid: Government by the People*. Published by Delphi Press in 1981, it analyzed the dysfunctional concepts and practices of bureaucratic government and posited self-renewing principles more appropriate for democratic institutions. From 1987 until 1995 his column "Global Perspective" was a regular feature of the newspaper *Washington International*. In addition, he has contributed to other books and published professional articles in his various fields of research over the years. He has produced a number of educational videos in the area of frontier science/social change.

Paul's graduate degrees in public administration and psychology are from Harvard University (MPA in 1974) and Florida State University (M.S. in 1962), from which he also received his B.A. (Phi Beta Kappa) in 1961. From 1962 to 1965, he was on active duty in the U.S. Navy as a line officer, continuing for several years as an intelligence officer in the active reserves. As a U.S. Foreign Service Officer in the Department of State from 1965 until 1979, Paul served in diplomatic posts in France, Martinique, Sierra Leone, and the Dominican Republic. He also served in various management and policy positions in Washington, D.C. during this period. Paul was the founding chief executive officer of the Washington-based nonprofit Delphi International from 1979 until 1995. In this capacity he designed and directed a large number of international educational exchange, training, and development programs involving participants from over one hundred countries.

In this long career, Paul carried on an active program of cross-cultural research in the fields of psychology, anomalous human experiences, and frontier science. He currently resides in Ashland, Oregon where he does research and writes in the fields covered in the *Solarian Legacy*; he produces educational video and television programs in these areas and is an active speaker on these themes. He may be contacted at 679 Clay Street, Ashland, OR 97520 and by e-mail at coslog@mind.net.

To order additional books or to request a catalog …

Simply call us at the toll-free number below. For catalogs to be sent outside of the USA, please send $3.00 for postage and handling. Book orders must be prepaid: check, money order, international coupon, VISA, MasterCard, Discover Card, and American Express accepted.

To place your order, call toll-free: 1 (888) ORDER IT (673-3748). Orders only, please!

For information, or to mail or fax an order, contact:

OUGHTEN HOUSE PUBLICATIONS
PO Box 2008
Livermore, CA 94551
Phone: (510) 447-2332
Fax: (510) 447-2376
Toll-free (888) ORDER IT
e-mail: oughten@oughtenhouse.com
Internet: www.oughtenhouse.com